S0-BLI-370

Water Soluble
Poly-*N*-Vinylamides

Water Soluble Poly-*N*-Vinylamides

Synthesis and Physicochemical Properties

Yury E. Kirsh
Karpov Institute of Physical Chemistry, Moscow, Russia

JOHN WILEY & SONS

Chichester · New York · Weinheim · Brisbane · Singapore · Toronto

Chemistry Library

Copyright © 1998   John Wiley & Sons Ltd,
Baffins Lane, Chichester,
West Sussex PO19 1UD, England

National   01243 779777
International (+44) 1243 779777
e-mail (for orders and customer service enquiries): cs-books@wiley.co.uk
Visit our Home Page on http://www.wiley.co.uk
or http://www.wiley.com

All rights Reserved. No part of this publication may be reproduced, stored in a
retrieval system, or transmitted, in any form or by any means, electronic, mechanical, photocopying,
recording, scanning or otherwise, except under the terms of the Copyright Designs and Patents Act 1988
or under the terms of a licence issued by the Copyright Licensing Agency, 90 Tottenham Court Road,
London W1P 9HE, UK, without the permission in writing of the Publisher.

Other Wiley Editorial Offices

John Wiley & Sons, Inc., 605 Third Avenue,
New York, NY 10158-0012, USA

WILEY-VCH Verlag–GmbH, Pappelallee 3,
D-69469 Weinheim, Germany

Jacaranda Wiley Ltd, 33 Park Road, Milton,
Queensland 4064, Australia

John Wiley & Sons (Asia) Pte Ltd, Clementi Loop #02-01,
Jin Xing Distripark, Singapore 129809

John Wiley & Sons (Canada) Ltd, 22 Worcester Road,
Rexdale, Ontario M9W 1L1, Canada

Library of Congress Cataloging-in-Publication Data

Kirsh, Yury E.
Water soluble poly-N-vinylamides : synthesis and physicochemical
properties / Yury E. Kirsh.
p.   cm.
Includes bibliographical references (p.   –   ) and index.
ISBN 0-471-97630-X (alk. paper)
1. Polyvinylamides.   I. Title.
QD383.V56K57   1998
547'.042—dc21   97-17696
CIP

British Library Cataloguing in Publication Data

A catalogue record for this book is available from the British Library

ISBN 0 471 97630 X

Typeset in 10/12pt Times by Thomson Press (India) Ltd, New Delhi
Printed and bound in Great Britain by Antony Rowe Ltd., Chippenham, Wilts.
This book is printed on acid-free paper responsibly manufactured from sustainable forestry, in which at
least two trees are planted for each one used for paper production

QD383
V56
K57
1998
CHEM

Contents

Preface

A generation of water-soluble carbon chain polymers involving amide goups in side substitutes, namely poly-*N*-vinylamides is represented in this book. The best-known representative of this generation of polymers is poly-*N*-vinylpyrrolidone (PVP). The interest in poly-*N*-vinylamides is caused by the large-scale that they give for the study of the relationship between the chemical structure of side substituents and their physicochemical properties in aqueous solutions. The synthetic approaches developed have allowed the substituent structure to be varied at amide groups.

PVP heads the list of poly-*N*-vinylamides. First developed in Germany by W. Reppe and his colleagues during the 1930s, PVP was used as a blood-plasma substitute during the Second World War. The synthesis of PVP was patented by W. Reppe and H. Fikentsher in 1939.

In Russia the first monograph on PVP entitled *Chemistry of N-Vinylpyrrolidone and its Polymers*, by F. P. Sidelkovskaya, was published in 1970. An extensive list of references concerning synthetic procedures of *N*-vinylpyrrolidone (VP) preparation, VP polymerization and physicochemical properties of PVP has been given in the book. It should be noted that as a result there has been a marked acceleration of PVP investigations in Russia.

In recent years the field of polymer chemistry has been actively developed. There are numerous scientific and experimental findings on PVP and other poly-*N*-vinylamides scattered in different references. Our own investigations over the last 20 years in this field and analysis of reference data has led us to believe in the great practical importance of the poly-*N*-vinylamides. Of the large numbers of water-soluble polymers these deserve great attention as models of natural polyamides (proteins). Polymers, especially PVP, find widespread application today in medicine, pharmaceuticals, cosmetics, foods, textiles, etc. PVP and copolymers of VP are found in the composition of various medical formulations, being carriers of biologically active compounds. PVP derivatives comprise sorbents, protective colloids, modifiers of membrane surfaces and solvents of drugs and dyes. In the future, the list of practical applications will be increased considerably. In practice, it is evident that poly-N-vinylamides with new attractive properties will be found.

A remarkable example is poly-N-vinylcamprolactam (PVCL) possessing a low critical solution temperature in aqueous solutions for a range of physiological temperatures (32–35 °C). This property of PVCL opens promising perspectives in the development of new approaches in immobilizing enzymes, cells and drugs.

It is hoped that both PVP derivatives and other poly-N-vinylamides with different structures of substituents will help to develop new directions in medicine, biotechnology, microbiology, membrane technology and so on.

Development of simple synthetic pathways of preparing N-vinylamide monomers and regulation of their polymerization processes are the keys to progress in the chemistry of water-soluble polymers. Knowledge of the features of conformational transformations within the macromolecules, their hydration and complexation with small molecules and various polymer compounds in aqueous solutions will add much to our concepts of water-soluble polymers in physical chemistry, and also natural macromolecules. I hope that this book represents one of the steps in this field of polymer chemistry.

The work is dedicated to the memory of Prof. K. S. Lyalikov under whose direction the author began his scientific activity at Leningrad Institute of Cinema Engineers (St. Petersburg). Being a scientist of the old Petersburg physicochemical school in a branch of photographic processes and photo materials, he predetermined my future scientific fate.

I am sincerely grateful to Prof. V. A. Kabanov (Moscow State University) who induced me to devote myself to chemistry and physico-chemistry of water-soluble polymers.

A significant part of the experimental research, the results of which are given in this book, was carried out in co-operation with Doctor of Sciences T. M. Karaputadze, Ph.D., T. A. Soos, A. I. Aksenov (Institute of Blood Substituents and Hormone Preparations in Moscow), Doctor of Sciences G. V. Shatalov (Voronezh State University), N. A. Yanul (Karpov Institute of Physical Chemistry in Moscow) and others to whom I express sincere gratitude.

This work would not have been possible without the fruitful co-operation of Prof. E. V. Anufrieva, Prof. M. G. Krakovyak (Institute of High Molecular Compounds of Russian Academy of Science in St. Petersburg), Prof. A. M. Vasserman (Institute of Chemical Physics in Moscow) and Prof. S. F. Timashev (Karpov Institute of Physical Chemistry).

Our contacts with Prof. K. K. Kalninsh (Institute of High Molecular Compounds of Russian Academy of Science) were especially useful over many years. He has calculated electronic states and structures of a series of N-vinylamides.

I also thank my son A. Vengerov for participation in the translation of the book into English.

It is hoped that this book will be of assistance to much research in developing new directions in medicine, biochemistry, biotechnology and in many other branches of chemistry.

Introduction

In a number of water-soluble polymers PVP occupies a special place. It has attracted the constant attention of researchers over the last 55 years. Its commercial success is owing to its film-forming and adhesive characteristics, unusual complexing ability, relatively inert behaviour towards salts and acids, its resistance to thermal degradation in water, biological compatibility and low toxicity. In other words, PVP possesses a unique combination of chemical, physicochemical and biological properties.

In fact it dissolves very well in water at any composition. Amide groups of PVP are inert to hydrolysis at heat treatment of its aqueous solutions up to 120–130 °C and to the influence of weak acids and bases in water, unlike N-methylpyrrolidone as an analogue of a PVP chain link. The PVP macromolecules in water form complexes involving molecules of different structure (inorganic and organic ions, polyelectrolytes and others). This list of the advantages of PVP is far from complete.

However, in spite of the attention to PVP science and its practical application, physicochemical fundamentals for the appearance of such a combination of properties remain to be explained. Each of these properties seem to result from the action of a series of numerous factors (structural, solvation, conformational and others) shown in water-dissolved PVP chains. Understanding of these factors will allow us to establish new fields of practical application of N-vinylamides and display new specific properties of these polymers.

Correlation and generalization of data concerning physicochemical properties of PVP, together with poly-N-vinylamides of different structure, will favour the understanding of the above-mentioned factors. Also considered in the book is a combination between conformational states of macromolecules in solution and structural transformations of solvent molecules, especially water molecules surrounding the chain links. Such an approach will give an insight into structural properties of the unique liquid which is water.

Since poly-N-vinylamides as a new generation of water-soluble polymers are of great interest in science and medical practice, it is necessary to combine synthesis of N-vinylamides, their conformations and electronic states in solutions, polymerization processes with their participation and physicochemical properties of poly-N-vinylamides shown in aqueous solutions.

In Chapter 1, the reader is introduced to N-vinylamides of various structures, common pathways of synthesis, conformations for side substituents in N-vinylamides of aliphatic carboxylic acids and rings in N-vinyllactams found by various physicochemical methods (^{1}H and ^{13}C NMR, X-ray analysis and quantum-chemical approaches), and medium effects on electronic states involving charge distribution on monomer atoms calculated by quantum-chemical methods.

Chapter 2 presents polymerization properties of N-vinylamides in different solvents, mechanism of radical polymerization of N-vinylpyrrolidone (VP) in a traditional system (VP, H_2O_2, NH_3, H_2O) and factors controlling polymer molecular weights in polymerization reaction, conformations of main chain and side residues in a series of N-vinylamides. Also given in this chapter are some non-traditional polymerization processes, in particular radical polymerization with concurrent opening VP ring polycondensation and reactivity ratio of monomers in copolymerization with a large group of monomers. Marked attention is given to the approaches permitting the determination of molecular weights of PVP and other polymers.

In Chapter 3, there are description of hydration phenomena in poly-N-vinylamide chains being detected by various physicochemical methods (^{1}H and ^{13}C NMR, IR spectroscopy, EPR of spin-labelled polymers, DSC, etc.). Attention is focussed on water transitions into polymer solutions (ice melting) and gels (ice melting and evaporation). The concept concerning structural transformations of water associates under the action of polar $C=O$ groups on poly-N-vinylamide chains in aqueous solution is developed. The partially negatively charged ends of $C=O$ dipoles strongly polarize water molecules near to the polyamide chain, changing the charge distribution on their atoms and lowering the number of hydrogen bonds.

Chapter 4 is a broad consideration of complex formations of PVP and other poly-N-vinylamides interacting with inorganic anions, dyes, synthetic and natural macromolecules and physicochemical principles of the interaction.

A series of reactions with participation of these polymers (hydrolysis, modification reactions of VP copolymers containing reactive groups, reaction of PVP and iodine in solution and in powder, ionization equilibrium of amine groups on poly-N-vinylamide chains, etc.) is given in Chapter 5. Aspects considered are those concerning the formation of ultrafiltration membranes by immersion of a polymer cast solution containing PVP in water and the processes occurring during the phase separation of a membrane-forming polymer + PVP system, the operation mechanism of reverse osmosis (RO) polyamide membranes through which these membranes function using new data obtained by studying poly-N-vinylamide hydration.

1

N-vinylamides: Synthesis, Conformation, and Electronic State

1.1 SYNTHESIS OF N-VINYLLACTAMS AND N-VINYLAMIDES OF ALIPHATIC CARBOXYLIC ACIDS

N-VINYLPYRROLIDONE

For N-vinyllactam preparation a large number of reactions is reported [1–3]. These can be divided into two groups: indirect vinylation and direct vinylation of lactams with acetylene.

To the first group it is necessary to attribute the reaction of interaction of γ-butyrolactone (I) and ethanolamine (II). The process includes substitution of the hydroxy group of compound (III) by chlorine atom (IV), followed by hydrogen chloride splitting off in the presence of the base, resulting in formation of VP [4].

$$
\begin{array}{ccc}
\underset{\substack{| \\ \text{CH}_2-\text{CH}_2}}{\overset{\substack{\text{O} \\ \diagup \diagdown}}{\text{CH}_2}} \text{C}{=}\text{O} & + \ \text{NH}-\text{CH}_2-\text{CH}_2-\text{OH} \ \longrightarrow & \quad (1.1) \\
\text{I} & \text{II}
\end{array}
$$

$$
\begin{array}{ccccc}
\text{CH}_2-\text{CH}_2-\text{OH} & & \text{CH}_2-\text{CH}_2-\text{Cl} & & \text{CH}{=}\text{CH}_2 \\
| & & | & & | \\
\text{N} & \xrightarrow{\text{SOCl}_2} & \text{N} & \xrightarrow[-\text{HCl}]{\text{KOH}} & \text{N} \\
\diagup \diagdown & & \diagup \diagdown & & \diagup \diagdown \\
\text{CH}_2 \quad \text{C}{=}\text{O} & & \text{CH}_2 \quad \text{C}{=}\text{O} & & \text{CH}_2 \quad \text{C}{=}\text{O} \\
| \quad\quad | & & | \quad\quad | & & | \quad\quad | \\
\text{CH}_2-\text{CH}_2 & & \text{CH}_2-\text{CH}_2 & & \text{CH}_2-\text{CH}_2 \\
\text{III} & & \text{IV} & & \text{VP}
\end{array}
$$

VP was also synthesized by reaction of *N*-(*β*-hydroxyethyl)pyrrolidone (III) with acetic anhydride and pyrolysis (V) [5]:

$$
\begin{array}{ccc}
\text{V} & \xrightarrow{460\ °C} & \text{VP}
\end{array}
\tag{1.2}
$$

Pyrolysis of compound (V) in the presence of activated aluminium oxide (93% Al_2O_3, 2% Fe_2O_3, 5% KOH) allowed the preparation of VP with high yields (82%) [6].

Reaction of α-pyrrolidone (VI) with vinylchloride [7] or vinylbutyl ether was also offered [8]:

$$
\text{VI} + CH_2{=}CH{-}Cl \longrightarrow \text{VP} + HCl
\tag{1.3}
$$

$$
\text{VI} + CH_2{=}CH{-}O{-}C_4H_9 \longrightarrow \text{VP} + C_4H_9{-}OH
\tag{1.4}
$$

Reppe's reaction of acetylene with lactams such as pyrrolidone or caprolactam found the greatest application in laboratory practice and in industry. It is based on interaction between lactam and acetylene at 130–180 °C mediated by alkaline catalysts, which are predominantly potassium lactam salts [1, 3].

$$
\text{VI} + CH{\equiv}CH \longrightarrow \text{VP}
\tag{1.5}
$$

The reaction is carried out both in solvents (toluene, dioxane or other) and in the absence of solvent under high pressure (~25 atm) at 100–105 °C.

Lactam conversion in the reaction is 50–60%. After removal of the unreacted amide, product yield reaches 80–83%. The reaction is described in a large number of patents which are reviewed in refs [2, 3]. However, the preparation technology of VP with acetylene under high pressure is hazardous because of the danger of explosion [9].

Attention of researchers was focussed on an increase of monomer yield, simplification of the process and VP quality improvement. An alternative method of synthesis with minimum explosion hazard was developed in Russia on the basis of research carried out by M. F. Shostakovsky and colleagues in the Institute of Organic Chemistry in Moscow [3, 10]. The process of VP synthesis was realized at acetylene pressure close to atmospheric in a vinylation reactor supplied with an electromagnetic drive of the mixing device, ensuring safe airtight operation to avoid possible formation of dangerously explosive acetylene–air mixtures [31].

A reaction mixture was prepared from pyrrolidone with the addition of 4–5% KOH (solid state or ethanol solution). After KOH dissolving in pyrrolidone, water or ethanol were removed on a film evaporator to prepare the catalyst (potassium pyrrolidonate salt).

The mixture was introduced into a vinylation reactor in which intensive agitation of gas (acetylene) and liquid was carried out at a Reynolds number equal to 40 000 with admission of purified and dried acetylene at a rate ensuring the maintenance of low pressure (0.5 bar) within the reactor. Vinylation reaction was carried out at 170 ± 5 °C at the continuously working mixing device and continuous accumulation of unreacted gas (acetylene). The reaction time takes 3–4 h, allowing a 50–60% conversion of pyrrolidone to VP to be carried out. At a longer time of reaction intensive formation of low and high molecular weight by-products takes place thereby reducing VP yield and polluting the final product. The reaction product prepared by this procedure is distilled to isolate VP raw material, followed by VP purification to pure monomer. As a result of this process, VP with high contents of ground substance (99.5–99.8%) was obtained.

$$
\begin{array}{c}
\underset{\underset{\text{VI}}{\overset{\displaystyle CH_2-CH_2}{|}}}{\overset{\displaystyle CH_2 \quad C=O}{|}}\overset{\displaystyle \diagup NH}{\diagdown} + KOH \longrightarrow \underset{\overset{\displaystyle CH_2-CH_2}{|}}{\overset{\displaystyle CH_2 \quad C=O}{|}}\overset{\displaystyle \diagup \bar{N}K^+}{\diagdown} + H_2O \qquad (1.6)
\end{array}
$$

However, it was seen [11–14] that some impurities of complicated structure were found in VP samples. The impurities combine with the monomer at the fractionating process and influence the subsequent polymerization reaction [12], and increases the pH value in aqueous monomer solution up to 9.5–10.5 (10 wt% VP) depending on VP quality. This signifies the presence of impurities containing amine groups in VP samples, as pure VP does not, in practice, change the pH of water [13].

Special attention was given to VP quality as the presence of impurities influences VP polymerization in water in the presence of hydrogen peroxide and ammonia, reducing polymer yield and increasing the colour index of the product [11, 13].

The appearance of amine by-products is likely to be due to rigid conditions (high temperature and long time of reaction) of vinylation reaction [3]. One of the possible reactions is hydrolysis of pyrrolidone caused by trace amounts of water in the presence of KOH with formation of γ-aminobutyric acid salt (VII) (1.7).

$$
\underset{\text{VI}}{\overset{\displaystyle \overset{\text{NH}}{\underset{\text{CH}_2-\text{CH}_2}{\text{CH}_2\quad\text{C}=\text{O}}}}{}} + \text{H}_2\text{O} \quad \xrightarrow{\text{K}^+} \quad \underset{\text{VII}}{\text{NH}_2-\text{CH}_2-\text{CH}_2-\text{CH}_2-\text{COO}^-\text{K}^+} \qquad (1.7)
$$

As is known, high temperature [15] causes decarboxylation of salts of organic acids with formation of primary amine. Thus compound VII can probably be decarboxylized to an amine compound VIII.

$$
\underset{\text{VII}}{\text{NH}_2-(\text{CH}_2)_3-\text{COO}^-\text{K}^+} \quad \xrightarrow{\overset{\text{NH}}{\underset{\text{CH}_2-\text{CH}_2}{\text{CH}_2\quad\text{C}=\text{O}}}}
$$

$$
\underset{\text{VIII}}{\text{NH}_2-\text{CH}_2-\text{CH}_2-\text{CH}_3} + \text{CO}_2 + \underset{\text{CH}_2-\text{CH}_2}{\overset{\text{N}^-\ \text{K}^+}{\text{CH}_2\quad\text{C}=\text{O}}} \qquad (1.8)
$$

An alternative possible reaction is the polymerization of cyclic amides initiated by base with formation of polyamides [16] with terminal amino groups (1.9).

$$
\underset{\text{VI}}{\overset{\text{N}^-\ \text{K}^+}{\underset{\text{CH}_2-\text{CH}_2}{\text{CH}_2\quad\text{C}=\text{O}}}} + \underset{\text{VI}}{\overset{\text{NH}}{\underset{\text{CH}_2-\text{CH}_2}{\text{CH}_2\quad\text{C}=\text{O}}}} \longrightarrow \underset{\text{IX}}{\overset{\text{O}}{\underset{\text{CH}_2-\text{CH}_2}{\text{CH}_2-\text{C}}}\overset{}{\underset{}{\text{N}-\overset{\text{O}}{\text{C}}-(\text{CH}_2)_3-\text{NH}^-\text{K}^+}}} \xrightarrow{\text{VI}}
$$

$$
\underset{X}{\overset{\text{O}}{\underset{\text{CH}_2-\text{CH}_2}{\text{CH}_2-\text{C}}}\overset{}{\underset{}{\text{N}-\overset{\text{O}}{\text{C}}-(\text{CH}_2)_3-\text{NH}-\overset{\text{O}}{\text{C}}-(\text{CH}_2)_3-\text{NH}^-\text{K}^+}}} \xrightarrow{\text{VI}}
$$

$$
\underset{XI}{\overset{\text{O}}{\underset{\text{CH}_2-\text{CH}_2}{\text{CH}_2-\text{C}}}\overset{}{\underset{}{\text{N}-\overset{\text{O}}{\text{C}}-(\text{CH}_2)_3-\text{NH}-\overset{\text{O}}{\text{C}}-(\text{CH}_2)_3-\text{NH}_2}}} + \underset{\text{CH}_2-\text{CH}_2}{\overset{\text{N}^-\ \text{K}^+}{\text{CH}_2\quad\text{C}=\text{O}}} \qquad (1.9)
$$

In the presence of a trace amount of water hydrolysis of polyamide (XI) is possible with formation of amine derivatives of complicated structure. Since pyrrolidone (VI) is synthesized in a reaction between butyrolactone (I) and ammonia (1.10) [2], the butyrolactone can be present in pyrrolidone as an impurity that reacts with extrinsic amines with formation of, e.g. N-propyl-pyrrolidone (XII) (reaction 1.11). In fact, a GLC chromatogram of VP obtained by vinylation reaction usually shows a peak near the VP peak which is related to propyl- (or ethyl-) pyrrolidone [13].

$$
\underset{\text{I}}{\begin{array}{c}\text{O} \\ \diagup \quad \diagdown \\ \text{CH}_2 \quad \text{C}{=}\text{O} \\ | \qquad | \\ \text{CH}_2{-}\text{CH}_2 \end{array}} + \text{NH}_3 \longrightarrow \text{HO}{-}(\text{CH}_2)_3{-}\text{CONH}_2 \longrightarrow \underset{\text{VI}}{\begin{array}{c}\text{NH} \\ \diagup \quad \diagdown \\ \text{CH}_2 \quad \text{C}{=}\text{O} \\ | \qquad | \\ \text{CH}_2{-}\text{CH}_2 \end{array}} \qquad (1.10)
$$

$$
\underset{\text{VIII}}{\text{CH}_3{-}\text{CH}_2{-}\text{CH}_2{-}\text{NH}_2} + \underset{\text{I}}{\begin{array}{c}\text{O} \\ \diagup \quad \diagdown \\ \text{CH}_2 \quad \text{C}{=}\text{O} \\ | \qquad | \\ \text{CH}_2{-}\text{CH}_2 \end{array}} \longrightarrow \underset{\text{XII}}{\begin{array}{c}\text{CH}_2{-}\text{CH}_2{-}\text{CH}_3 \\ | \\ \text{N} \\ \diagup \quad \diagdown \\ \text{CH}_2 \quad \text{C}{=}\text{O} \\ | \qquad | \\ \text{CH}_2{-}\text{CH}_2 \end{array}} \qquad (1.11)
$$

It is also impossible to rule out the reaction involving interaction of acetylene with amine groups of the impurities. As specified in ref. [2], vinylation of amines, e.g. ethanolamine, by acetylene can be accompanied by formation of small amounts (about 1%) of unsaturable amines (XIII, XIV).

$$
\underset{\text{XIII}}{\text{CH}_2{=}\text{CH}{-}\text{NH}{-}\text{CH}_2{-}\text{CH}_2{-}\text{CH}_3} \quad \text{or} \quad \underset{\text{XIV}}{(\text{CH}_2{=}\text{CH})_2\text{N}{-}\text{CH}_2{-}\text{CH}_2{-}\text{CH}_3}
$$

In this case the combination of secondary reactions can form more complicated unsaturable compounds, for example heterocyclic ones.

Of course, most of the impurities are removed by rectification. However, a series of external amines is apparently present in VP steam then remaining in monomer. That is why a number of VP chemical treatment procedures was offered for its removal [13, 14] involving diisocyanate reaction [13] (reaction 1.12), condensation with polyacrylic acid [13] and weak acidic ion-exchange resin treatment [14].

$$
\text{R}_1{-}\text{NH}_2 + \text{O}{=}\text{C}{=}\text{N}{-}\text{R}_2{-}\text{N}{=}\text{C}{=}\text{O} \longrightarrow
$$

$$
\longrightarrow \text{R}_1{-}\text{NH}{-}\overset{\overset{\text{O}}{\|}}{\text{C}}{-}\text{NH}{-}\text{R}_2{-}\text{NH}{-}\overset{\overset{\text{O}}{\|}}{\text{C}}{-}\text{NH}{-}\text{R}_1 \qquad (1.12)
$$

The fact that an addition of small amounts of toluylenediisocyanate (0.01–0.1%) to the monomer, followed by distillation, permits preparation of a pure product which does not change the value of pH in water, indicates the presence of amines of unknown structure. The preferred contents of external amines on a background of other impurities in purified VP found confirmation in pH reduction (from 9.5 to 7.0) of 10% aqueous solution of initial VP at passage through a cation-exchange resin in H⁺-form (amberlite) [14].

Single-stage procedure of VP synthesis from acetylene and pyrrolidone used in industry seems to be simple. However, it possesses a series of shortcomings. Rigid conditions of reaction cause formation of a number of by-products of complicated structure, poorly removed from the monomer and requiring an additional purification stage. As the vinylation reaction does not proceed completely (yield 50–60%), the residual unreacted pyrrolidone is rectified, purified and distillated for the preparation of pure product which re-enters the reaction.

Therefore, there is an obvious necessity of development of an alternative VP synthesis procedure allowing the preparation of high-quality VP without additional processing.

N-VINYLCAPROLACTAM

The basic characteristics of the caprolactam vinylation process under acetylene pressure close to atmospheric are described in ref. [3]. It was found that the potassium salt of carpolactam is a more effective catalyst for vinylation reaction in comparison with sodium salt. So at the same amount (~ 7 wt%) of catalyst (sodium caprolactam or potassium caprolactam) the *N*-vinylcaprolactam (VCL) content in a reactive mixture after performing vinylation reaction during 1 h at 135 °C is 54.8 wt% in the presence of potassium caprolactam against 13.3 wt% in the presence of sodium caprolactam [3].

The intensity of mixing defines the vinylation reaction rate [3]. The optimum rotation rate of a turbo-stirrer is in the range from 3000 to 5000 rpm to create a gas–liquid emulsion of acetylene in caprolactam at a low gas pressure. Optimum reaction time is 3 h. During reaction, accumulation of polymeric viscous products occurs which reduces monomer yield due to caprolactam concentration decrease and pollutes the monomer by collateral products. Caprolactam in the form of the base catalyst probably enters a known anionic polymerization reaction (reaction 1.9) [16].

Activation energy (ΔH) of vinylation reaction is ~ 54.4 kJ/mole. The reaction rate constant is $2.5 \times 10^{-3}\,\mathrm{s}^{-1}$ at 135 °C [3].

The two-stage distillation of the reaction mixture is conducted to prepare pure VCL. At the first stage VCL raw material is obtained after its rectification at 86–126 °C (2–10 mmHg) and separation of VCL and unreacted caprolactam from resinification products. At the second stage a distillation purification of VCL raw material at 82–84 °C (1.5–3 mmHg) is conducted.

As the vinylation reaction of caprolactam is realized in softer conditions (130–140 °C) than in the case of pyrrolidone, the monomer does not contain such a large quantity of by-products as VP.

N-VINYLAMIDES OF ALIPHATIC CARBOXYLIC ACIDS

These compounds were synthesized by direct vinylation [17–19].

$$
\begin{array}{c}
R_2 \diagdown \\
\quad C{=}O \\
\quad | \\
\quad NH \\
R_1 \diagup
\end{array}
\; + \; CH{\equiv}CH \; \longrightarrow \;
\begin{array}{c}
R_2 \diagdown \\
\quad C{=}O \\
\quad | \\
\quad NH{-}CH{-}CH_2 \\
R_1 \diagup
\end{array}
\qquad (1.13)
$$

In the liquid phase this reaction was conducted at 100–200 °C using such alkali metal-based catalysis as, for example, sodium, metallic sodium or potassium amides [18]. When conducting the reaction in the gas phase the catalysts used were potassium vanadate, potassium zirconate, potassium tungstate, etc. However, in this case the monomer yield was lower, reaching only 10–20% in N-vinyl-N-methylacetamide for example [20]. The evolution of gaseous products was reported when acetylene was purged through amides in the presence of alkaline catalysts. This drawbacks to direct vinylation compelled the researchers to look for other, more simple and economical synthetic methods for the preparation of vinylamides.

Blauche, Langmeadow and Cohen [21] devised a procedure for N-vinyl-N-methylacetamide preparation using Schiff's bases. The addition of acetaldehyde into a solution of methylamine with alkali (NaOH) cooled to – 20 °C resulted in the formation of N-ethylidenemethylamine with 80% yield (1.14).

$$
CH_3{-}C\!\!\underset{H}{\overset{O}{\diagup}}\; + \; CH_3{-}NH_2 \; \longrightarrow \; CH_3{-}CH{=}N{-}CH_3 \qquad (1.14)
$$

A 1 : 1 mixture of triethylamine with acetic anhydride was gradually added to the Schiff base at – 10 °C. The reaction mixture was kept at 0 °C for two or three days and then heated for five hours. Acetic acid was neutralized with Na_2CO_3 in water at 0 °C (1.15).

$$
CH_3{-}CH{=}N{-}CH_3 \; + \; (CH{-}CO_2)_3O \; \longrightarrow \;
\begin{array}{c}
O{-}COCH_3 \\
| \\
CH_3{-}CH{-}N{-}CH_3 \\
| \\
CH_3{-}C{=}O
\end{array}
\; \longrightarrow
$$

$$
\xrightarrow{(C_2H_5)_3N} \;
\begin{array}{c}
CH_2{=}CH{-}N{-}CH_3 \\
| \\
CH_3{-}C{=}O
\end{array}
\; + \; CH_3{-}C\!\!\underset{OH}{\overset{O}{\diagup}} \qquad (1.15)
$$

The monomer was separated by extraction and distilled (yield 80%). The second stage of synthesis was also carried out by application of acetic acid, triethylamine and ketene [22].

A wide range of *N*-substituted vinylamides was synthesized in reactions of Schiff's bases with anhydrides of various carboxylic acids [23].

$$R_1-N{=}CH-CH_3 \ + \ (R_2CO)_2O \xrightarrow{\ (C_2H_5)_3N\ }$$

$$\text{(1.16)}$$

IR spectroscopy discovered intermediary *N*-acyloxyethylalkylamides of carboxylic acids when the base and the anhydride reacted (reaction 1.16). IR spectra of intermediates showed two intense bands with maxima at 1666 and 1740 cm^{-1}, characteristic of the C$=$O bond in the amide and ester groups respectively. Heating of the reaction mixtures leads to the abstraction of acid to form the corresponding *N*-vinyl-*N*-alkylamide. Table 1.1 represents the yield of the monomers and some of its physicochemical constants.

Bestian and Schnabel [24] conducted the condensation reaction of actaldehyde with various amides (1.17) and found an efficient method of the process under soft conditions.

N-methylformamide, *N*-methylacetamide, and other similar amides can be used as starting reagents. The suggested procedure involves heating of a mixture of acetaldehyde with amide at 20–70 °C in the presence of acidic of basic cata-

Table 1.1 Monomer yield (reaction 1.16) and physicochemical constants of *N*-vinyl-*N*-alkylamides [20]. Reproduced by permission of Nauka

R_1	R_2	Yield (%)	Temperature pressure (°C/kPa)	Extinction coefficient at $\lambda = 233$ nm (ethanol) (l mol^{-1} cm^{-1})	n_D^{20}
CH_3	CH_3	60	30–31/2.66	15 350	1.4830
C_2H_5	CH_3	68	65–70/1.99	15 260	1.4790
CH_3	C_2H_5	60	70–74/1.99	15 260	1.4755
C_2H_5	C_2H_5	71	55–60/0.66	15 370	1.4720
CH_3	C_3H_7	45	67–69/1.33	15 240	1.4710
C_2H_5	C_3H_7	56	78–82/0.66	15 450	1.4715
CH_3	C_4H_9	66	85–90/1.33	14 950	1.4705
C_2H_5	C_4H_9	68	61–63/0.26	15 100	1.4720

$$
R_2-NH \atop R_1-C=O \quad + \quad CH_3-C{\displaystyle{{\nearrow O}\atop{\searrow H}}} \quad \longrightarrow \quad R_2-N-\overset{\overset{\displaystyle OH}{|}}{C}H-CH_3 \atop R_1-C=O
$$

$$
\xrightarrow{-H_2O} \quad R_2-N-CH=CH_2 \atop R_1-C=O \qquad (1.17)
$$

lysts; the mixture is transferred to a thin-film evaporator at 180–250 °C at low pressure. Amide and vinylamide together are then distilled from the reaction mixture and are further separated by fractional distillation (1.17) [24, 25].

$$
NH_2-C{\displaystyle{{\nearrow O}\atop{\searrow H}}} + CH_3-C{\displaystyle{{\nearrow O}\atop{\searrow H}}} \longrightarrow CH_3-\overset{\overset{\displaystyle OH}{|}}{C}H-NH-C{\displaystyle{{\nearrow O}\atop{\searrow H}}}
$$

$$
\xrightarrow{NH_2-C{\nearrow O \atop \searrow H}} \quad CH_3-CH{\displaystyle{{\nearrow NH-C{\nearrow O \atop \searrow H}}\atop{\searrow NH-C{\nearrow O \atop \searrow H}}}} \qquad (1.18)
$$

In recent years special attention has been focussed on the reaction in search of optimum conditions of preparation, for example, N-vinylformamide (1.18, 1.19).

$$
CH_3-CH{\displaystyle{{\nearrow NH-C{\nearrow O \atop \searrow H}}\atop{\searrow NH-C{\nearrow O \atop \searrow H}}}} \xrightarrow{\text{pyrolysis}} CH_2=CH-NH-C{\displaystyle{{\nearrow O}\atop{\searrow H}}} + NH_2-C{\displaystyle{{\nearrow O}\atop{\searrow H}}}
$$

$$(1.19)$$

Dawson and Olleson [26] offered the condensation reaction of formamide in the presence of cation-exchange resin in H^+-form at 120 °C during 1–24 h, forming ethylene-bis-formamide (1.18). N-vinylformamide was prepared by pyrolysis of the latter (1.19). The same monomer was obtained by synthesis of N-(α-alkoxyethyl)formamide, followed by pyrolytic reaction [27] (1.20).

$$
NH_2-C{\displaystyle{{\nearrow O}\atop{\searrow H}}} + CH_3-C{\displaystyle{{\nearrow O}\atop{\searrow H}}} \longrightarrow CH_3-\overset{\overset{\displaystyle OH}{|}}{C}H-NH-C{\displaystyle{{\nearrow O}\atop{\searrow H}}} \xrightarrow{R_1OH}
$$

$$
CH_3-\overset{\overset{\displaystyle O-R_1}{|}}{C}H-NH-C{\displaystyle{{\nearrow O}\atop{\searrow H}}} \xrightarrow{\text{pyrolysis}} CH_2=CH-NH-C{\displaystyle{{\nearrow O}\atop{\searrow H}}} \qquad (1.20)
$$

As a catalyst of condensation, K_2CO_3 or K_3PO_4 in an amount of 0.1–10 mole% of the amide quantity and as a reagent (solvent), alcohols of different structure (primary, secondary [27] or polyatomic [28]) were introduced. High yield (95%) of the monomer was attained after pyrolytic elimination of alcohol from N-(α-alkoxyethyl)formamide [29].

Two procedures of purification were developed to obtain N-vinylformamide of high quality including fractional distillation [30] and monomer passing through a cation-exchange resin [31]. The problem of N-vinylformamide purification is also considered in patents [32, 33].

Brunnmuller et al. [34] described an original synthetic pathway of nitryl-formylamine thermal treatment at 250–600 °C (1.21) resulting in N-vinyl-formamide preparation.

$$HO-CH_2-NH-CH_2-CH_2-CN \xrightarrow{-HCN} \underset{H}{\overset{O}{\diagdown}}C-NH-CH=CH_2 \quad (1.21)$$

A series of approaches for N-vinylacetamide preparation was reported by Summerville and Stackman [35] concerning the condensation reaction between acetaldehyde and acetamide (1.22).

$$CH_3-C\overset{O}{\underset{H}{\diagup}} + 2\,CH_3-C\overset{O}{\underset{NH_2}{\diagup}} \longrightarrow CH_3-CH\overset{NH-\overset{O}{\overset{\|}{C}}-CH_3}{\underset{NH-\underset{O}{\underset{\|}{C}}-CH_3}{}} \quad (1.22)$$

The reaction proceeds in the presence of H_2SO_4 at an excess of acetamide forming ethylene-bis-acetamide, pyrolysis of which gives the required product [36].

$$CH_3-CH\overset{NH-\overset{O}{\overset{\|}{C}}-CH_3}{\underset{NH-\underset{O}{\underset{\|}{C}}-CH_3}{}} \xrightarrow{pyrolysis} CH_2=CH-NH-C\overset{O}{\underset{CH_3}{\diagup}} + NH_2-C\overset{O}{\underset{CH_3}{\diagup}} \quad (1.23)$$

An alternative process of synthesis is based on acetamide condensation involving dimethylacetal (1.24), with aftertreatment by decomposition within a pyrolytic column (1.25) [37, 38].

$$CH_3-CH\overset{OCH_3}{\underset{OCH_3}{\diagup}} + CH_3CONH_2 \xrightarrow{H^+} CH_3-CH\overset{NH-\overset{O}{\overset{\|}{C}}-CH_3}{\underset{OCH_3}{}} + CH_3OH \quad (1.24)$$

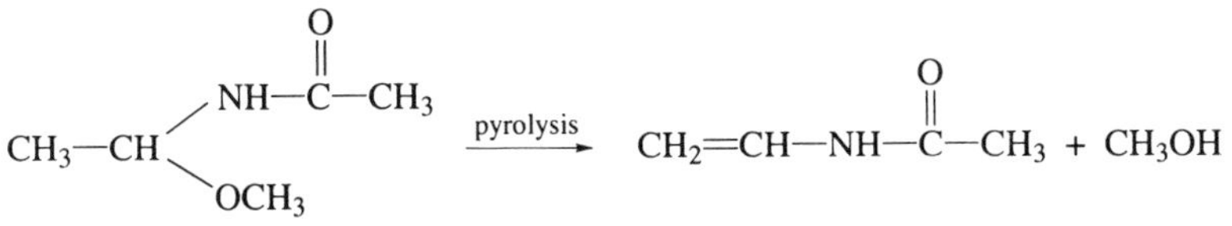

$$CH_3-CH \overset{NH-\overset{\displaystyle O}{\overset{\|}{C}}-CH_3}{\underset{OCH_3}{\diagup}} \xrightarrow{\text{pyrolysis}} CH_2=CH-NH-\overset{\displaystyle O}{\overset{\|}{C}}-CH_3 + CH_3OH \quad (1.25)$$

Akashi *et al.* [39] offered an application of methanol and acetaldehyde instead of dimethylacetal to increase monomer yield and quality (1.26, 1.27).

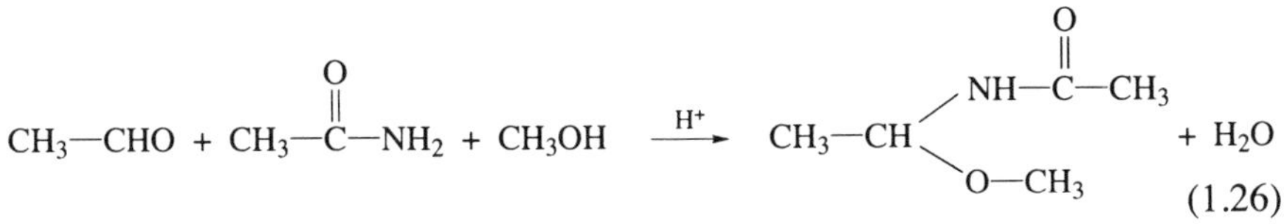

$$CH_3-CHO + CH_3-\overset{\displaystyle O}{\overset{\|}{C}}-NH_2 + CH_3OH \xrightarrow{H^+} CH_3-CH \overset{NH-\overset{\displaystyle O}{\overset{\|}{C}}-CH_3}{\underset{O-CH_3}{\diagup}} + H_2O \quad (1.26)$$

$$CH_3-CH \overset{NH-\overset{\displaystyle O}{\overset{\|}{C}}-CH_3}{\underset{O-CH_3}{\diagup}} \xrightarrow{\text{pyrolysis (500 °C)}} CH_2=CH-NH-\overset{\displaystyle O}{\overset{\|}{C}}-CH_3 + CH_3OH \quad (1.27)$$

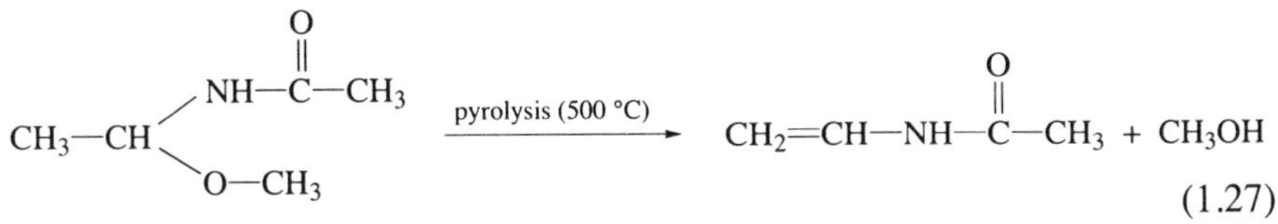

Etheramide can be isolated (yield 84%) and easily purified, allowing additional yield and quality of *N*-vinylacetamide.

Jensen, Schimdt and Mitralaff [40] described anodic alkoxylation of amides resulting in preparation of *N*-α-alkoxyethylamide of carboxylic acid (yield 40–50%).

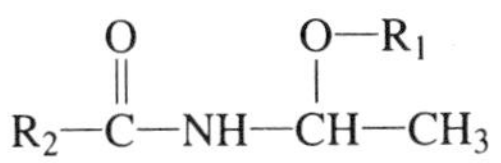

$$R_2-\overset{\displaystyle O}{\overset{\|}{C}}-NH-\overset{\overset{\displaystyle O-R_1}{|}}{CH}-CH_3$$

It was then alkylated with alkyl halides in alkaline medium in the following reaction (1.28):

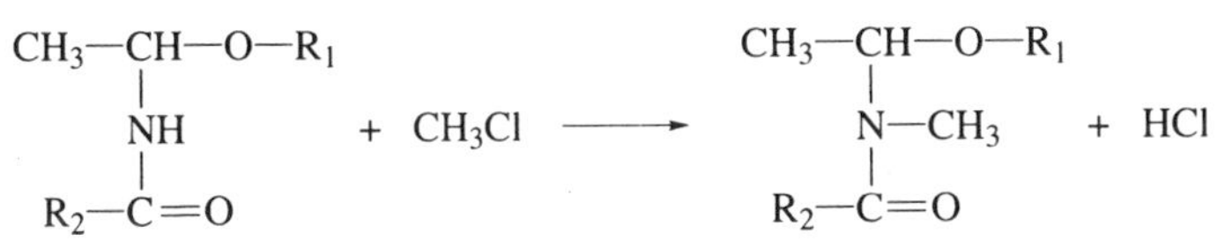

$$\begin{matrix} CH_3-CH-O-R_1 \\ | \\ NH \\ | \\ R_2-C=O \end{matrix} + CH_3Cl \longrightarrow \begin{matrix} CH_3-CH-O-R_1 \\ | \\ N-CH_3 \\ | \\ R_2-C=O \end{matrix} + HCl \quad (1.28)$$

The product in a nitrogen atmosphere was rapidly heat treated at 200 °C–300 °C with a catalyst (porous silicic acid), and the monomer was distilled with a final yield of 80–90% (1.29).

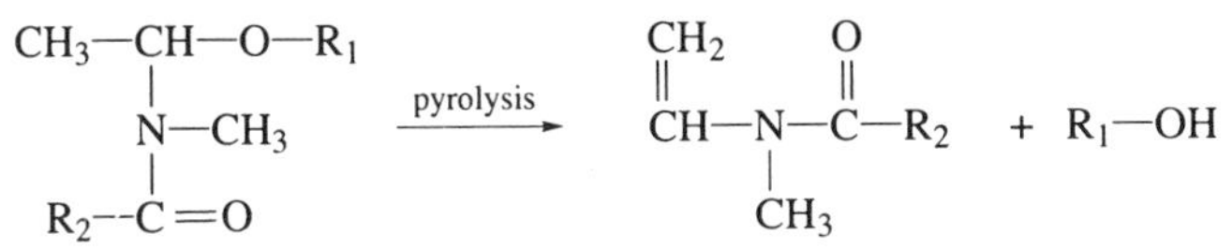

$$\begin{matrix} CH_3-CH-O-R_1 \\ | \\ N-CH_3 \\ | \\ R_2-C=O \end{matrix} \xrightarrow{\text{pyrolysis}} \begin{matrix} CH_2 \quad\quad O \\ \| \quad\quad \| \\ CH-N-C-R_2 \\ | \\ CH_3 \end{matrix} + R_1-OH \quad (1.29)$$

This procedure was used to fabricate *N*-vinyl-*N*-methylacetamide and *N*-vinyl-*N*-methylformamide.

Schwiersch and Hartwinner [41] proposed an alternative approach to the synthesis of *N*-vinylformamides with various alkyl substituents on nitrogen, e.g. *N*-vinyl-*N*-ethylformamide (VEF) or *N*-vinyl-*N*-propylformamide (VPF), which is based on the following chemical transformation (1.30).

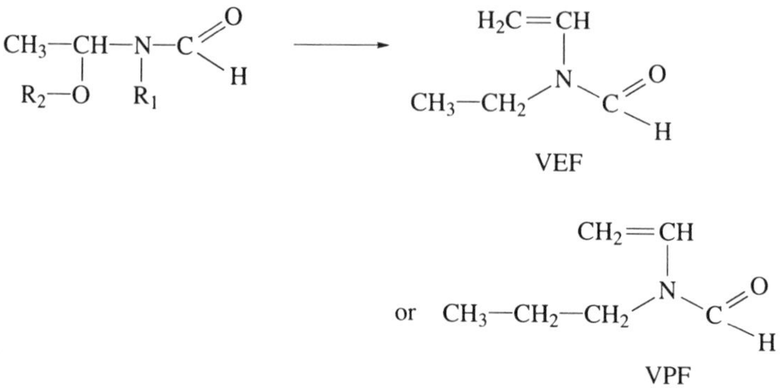

VEF

(1.30)

VPF

The pyrolytic reaction (1.30) is conducted at 100–200 °C under vacuum with weak acidic catalyst, e.g. alumina.

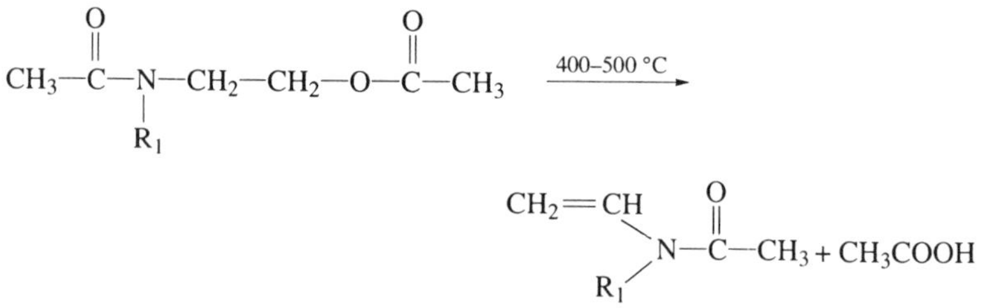

(1.31)

In this procedure [42], the initial reagents are synthesized from *N*-substituted ethanolamines and acetic anhydride and after pyrolysis (1.31) gave *N*-vinyl-*N*-alkylacetamides. The decomposition is accompanied by formation of acetic acid, which is removed by azeotropic distillation with an organic solvent, for example, xylene or dibutyl ether. Table 1.2 lists basic physical constants of *N*-vinyllactams and other *N*-vinylamides [20].

Thus, in this chapter an extensive body of synthetic pathways of preparation of *N*-vinylamides is represented. In fact, for *N*-vinyllactams the vinylation reaction is the most widespread. The approach is used in Germany (BASF), USA (International Specialty Products) and Russia. The technology has been used for a long period of time (40–50 years). Considerable technological know-how of *N*-vinylpyrrolidone preparation has accumulated during the last 10–15 years

Table 1.2 Basic physicochemical constants of N-vinylamides [20]. Reproduced by permission of Nauka

Monomer	Boiling temperature pressure (°C/kPa)	n_D^{20}	Ref.
N-vinylformamide	74/0.8		[27]
N-vinylacetamide	95/1.33		[35]
N-vinyl-N-methylformamide	45/1.87		[24]
N-vinyl-N-methylacetamide	64/2.67		[24]
	58/1.73	1.4791	[42]
	63.8/2.67	1.4827	[17]
	51.5/1.46	—	[40]
N-vinyl-N-ethylacetamide	57/1.6	1.4715	[42]
	62/1.73	—	[40]
	69/2.4	1.4780	[17]
N-vinyl-N-isopropylacetamide	74/2.67	1.4587	[42]
N-vinyl-N-methylpropionamide	69/1.73	1.4797	[17]
N-vinyl-N-ethylpropionamide	74/1.6	1.4737	[17]
N-vinylpyrrolidone	80/1.33	1.5110	[24]
N-vinylcaprolactam	95/0.53	1.5133	[2]

[12–14], and has considerably improved VP quality. However, studies on improving VP quality are continuing [13, 14].

The attention given to the VP monomer is limited by the fact that its impurities affect the following polymerization stages, decreasing polymer yield and causing sporadic colour index appearance. In addition, the high quality of VP polymers is defined by medical requirements since homo- and copolymers of VP enter into drug compositions [43]. Therefore, additional stages of purification of both VP [13, 14] and PVP aqueous solutions were proposed.

It should be noted that these stages complicate the technology, making the fabrication of VP and PVP more expensive. It is necessary to search for an alternative procedure which would be competitive with a known one concerning monomer yield and its quality.

Among N-vinylamides only the N-vinyl-N-methylacetamide is made on an industrial scale (Hoehst). The production of N-vinylformamide for laboratory use is carried out by Mitsubishi Chem. Ind., Ltd, Japan [44, 45], BASF, Germany [30], Air Products and Chemicals, USA [26]. Analysis of patents allows us to assume that the N-vinyl-N-methylacetamide industrial preparation procedure is based on a synthetic pathway with participation of Schiff's bases and acetic anhydride with subsequent pyrolysis [21]. The technology of N-vinylformamide is probably based on the condensation reaction of acetaldehyde and formamide [26, 35, 36].

1.2 CONFORMATIONS OF SIDE SUBSTITUENTS

CONFORMATIONAL STATE OF ACYCLIC N-VINYLAMIDES

Because of the high barrier for internal rotation about the $-\overset{\overset{\textstyle O}{\|}}{C}-N\overset{\diagup}{\diagdown}$ bond, low molecular weight *N*-substituted amides display two conformational states [46]:

trans-form   cis-form

The existence of conformers of *N*-vinylamides in solution was discovered by ^{13}C and ^{1}H NMR [47, 48]. The ^{1}H NMR spectra of *N*-vinylformamide, *N*-vinyl-*N*-methylformamide, and *N*-vinyl-*N*-methylacetamide show two signals associated with protons in the $H-\underset{|}{C}=$ group of double bonds (Fig. 1.1). It should be noted that one signal associated with the same groups was found in ^{1}H NMR spectra of *N*-vinylacetamide and, of course, in that of VP.

This indicates the presence of two stable conformers in the former and one in the latter. The presence of two conformers in the first three monomers is confirmed also by the fact that signals associated with aldehyde (*N*-vinylformamide and *N*-vinyl-*N*-methylformamide), and methyl (*N*-vinyl-*N*-methylformamide and *N*-vinyl-*N*-methylacetamide) protons are split into two peaks of different intensity. It is necessary also to note that in the case of *N*-vinylformamide an intensive signal of a proton in the methyne group is downfield compared with that of low intensity, the former being upfield and the latter being downfield for *N*-vinyl-*N*-methylformamide and *N*-vinyl-*N*-methylacetamide respectively. The different ratio of intensity of these signals in *N*-vinylformamide, on the one hand, and in *N*-vinyl-*N*-methylformamide with *N*-vinyl-*N*-methylacetamide, on the other hand, indicates its unequal contents in a series of *N*-vinylamides. In Table 1.3 the value of ^{1}H NMR chemical shifts of protons in monomers are investigated and also the ratio of splitting signals are represented.

As the assignment of each of the splitting peaks of a signal of a proton in $=$CH$-$ group to a certain type of conformer is inconvenient, the determination of the types was carried out by ^{13}C NMR spectra of *N*-vinylamides and VP with known arrangements of C$=$O and CH$_2$$=CH-$ groups (*trans*-form) (Table 1.4) [48].

It follows from these data that the signal of the carbon atom in $=\overset{*}{C}$H$-$ group of VP and of *N*-vinylacetamide has one peak, indicating the presence of one conformer (*trans*-form) with 100% by the contents. The signal of the same atom in *N*-vinylformamide (VF), *N*-vinyl-*N*-methylformamide (VMF), and *N*-vinyl-*N*-methylacetamide (VMA), is split into two peaks. In the case of the last two

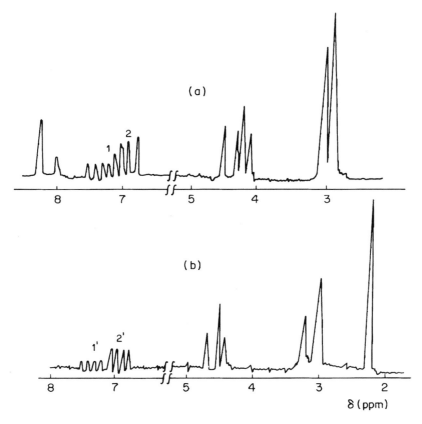

Fig. 1.1. ^{1}H NMR spectra of N-vinyl-N-methylformamide (a), and N-vinyl-N-methylacetamide (b) in D_2O (30%) [20]:

$$1,1' = \quad \underset{CH_2=CH}{\overset{CH_3}{\diagdown}}N-C\underset{O}{\overset{\diagup}{\diagdown}} \qquad 2,2' = \quad \underset{CH_2=CH}{\overset{CH_3}{\diagdown}}N-C\underset{\diagdown}{\overset{O}{\diagup}}$$

monomers the peak with high intensity is downfield, and the peak with low intensity is upfield. The intensity ratio of the peaks is 70 : 30 (Table 1.5). In the case of VF, the former appears upfield and latter downfield.

A comparison of chemical shifts (^{13}C NMR) in =CH— groups of VP (131 ppm), N-vinylacetamide (VAA) (one peak at 130.5 ppm), N-vinylformamide (VF) (high intensity peak at 130.4 ppm), and N-vinyl-N-methylacetamide (VMA) (low intensity peak at 131.2 ppm) (Table 1.4), shows close affinity of signal shift values which are to be referred to the *trans*-conformer [48]. In this case VAA, VF and VMA contain 100, 70 and 30% of the latter respectively.

Table 1.3 Chemical shifts (δ, ppm) ^{1}H NMR spectra of *N*-vinylamides (60 °C, D$_2$O) [47, 48]. The value of $\delta(H)$ for groups of $\overset{\diagdown}{\diagup}$N—$\overset{*}{C}H_2$—CH$_3$ (VEA), $\overset{\diagdown}{\diagup}$N—CH$_2$—$\overset{*}{C}H_2$ (VPA),

$\overset{\diagdown}{\diagup}$N—CH$_2$—CH$_2$—$\overset{*}{C}H_3$ (VPA), $-\overset{O}{\overset{\|}{C}}-CH_2-\overset{*}{C}H_3$ (VEP), and $-\overset{O}{\overset{\|}{C}}-\overset{*}{C}H_2-CH_3$ (VEP) corresponds to 1.34, 1.52, 0.86, 2.47 and 1.10 ppm, respectively. Accordingly, the ratio of intensities (per cent) of peaks is given in parentheses. Reproduced by permission of Nauka

No.	Amide	=CH—	CH$_2$=	$\overset{*}{C}H_3$—N$\overset{\diagup}{\diagdown}$ or $-\overset{*}{C}H_2$—N$\overset{\diagup}{\diagdown}$	$\overset{*}{C}H_3$—C$\overset{\diagup O}{\diagdown O}$ r	$-C\overset{\diagup O}{\diagdown_*^H}$
1.	VP	6.73	4.40	3.4	2.33	—
			4.45			
2.	VAA	6.57 (100%)	4.45		1.8	
			4.55			—
3.	VF	6.45 (30%)	5.25			
		6.67 (70%)				—
4.	VMF	6.90 (70%)	4.37	—		
		7.29 (30%)	4.52			
5.	VMA	7.04 (70%)	4.56	2.94 (70%)	—	—
		7.40 (30%)	4.68	3.05 (30%)		
6.	VEA	7.08 (70%)	4.50	3.07 (70%)	2.24	7.85
		7.44 (30%)	4.55	3.17 (30%)		
7.	VPA	6.86	4.30		2.45	8.16 (30%)
			4.45	—		8.39 (70%)
8.	VEP	6.94	4.29		2.16	
			4.47			

Notes: abbreviations: VP – *N*-vinylpyrrolidone, VAA – *N*-vinylacetamide, VF – *N*-vinylformamide, VMF – *N*-vinyl-*N*-methylformamide, VMA – *N*-vinyl-*N*-methylacetamide, VEA – *N*-vinyl-*N*-ethylacetamide, VPA – *N*-vinyl-*N*-propylacetamide and VEP – *N*-vinyl-*N*-ethylpropioneamide.

$$CH_2{=}CH\overset{\diagdown}{\underset{H}{N}}{-}C\overset{\diagup O}{\diagdown CH_3}$$

VAA (100% *trans*-form)

$$CH_2{=}CH\overset{\diagdown}{\underset{H}{N}}{-}C\overset{\diagup O}{\diagdown H} \rightleftharpoons CH_2{=}CH\overset{\diagdown}{\underset{H}{N}}{-}C\overset{\diagup H}{\diagdown O}$$

trans-form (70%)   *cis*-form (30%)

VF

$$CH_2{=}CH\diagdown N{-}C\diagup{\diagup}O \rightleftharpoons CH_2{=}CH\diagdown N{-}C\diagup CH_3$$

trans-form (30%)   cis-form (70%)

VMA

Table 1.4 Chemical shifts of carbon atoms (^{13}C NMR) in the monomers [47, 48]. The value of $\delta(^{13}C)$ of carbon in groups: $\diagdown N{-}CH_2{-}\overset{*}{C}H_3$ (VEA), $\diagdown N{-}CH_2{-}\overset{*}{C}H_3$ (VEP), $\diagdown N{-}CH_2{-}\overset{*}{C}H_2{-}CH_3$ (VPA), and $\diagdown N{-}CH_2{-}CH_2{-}\overset{*}{C}H_3$ (VPA) corresponds to 10.0, 10.6, 19.0 and 9.8 ppm respectively. Abbrevations are as Table 1.3. Reproduced by permission of Nauka

No.	Amide	=CH—	CH₂=	$CH_3{-}N\diagup$ or $-CH_2{-}N\diagup$	$\overset{*}{C}H_3{-}C\diagup{\diagup}O$ or $-\overset{*}{C}H_2{-}C\diagup{\diagup}O$	$-\overset{*}{C}\diagup{\diagup}O$
1.	VP	131.0	99.3	47.8	33.86	178.5
2.	VAA	130.5 (100%)	100.1	—	24.46	173.7
3.	VMA	131.2 (30%) 133.5 (70%)	91.1	26.5 (70%) 30.4 (30%)	20.0	167.3
4.	VEA	131.6 (70%) 129.2 (30%)	90.8	34.4 (70%) 37.6 (30%)	20.0	166.7
5.	VPA	132.4 (70%) 130.0 (30%)	91.4	41.6 (70%) 45.0 (30%)	20.4	167.3
6.	VEP	131.2	91.4	35.3	25.1	170.6

Table 1.5 Contents of cis-conformer in N-vinylamides of aliphatic carboxylic acids by ^{1}H and ^{13}C NMR [48]. Reproduced by permission of Nauka

		Fraction of cis-form (%)		
		D₂O		Pure compound
No.	Monomer	25 °C	60 °C	60 °C
1.	N-vinylformamide	30	30	30
2.	N-vinylacetamide	0	0	0
3.	N-vinyl-N-methylacetamide	70	69	70
4.	N-vinyl-N-ethylacetamide	68	65	68
5.	N-vinyl-N-propylacetamide	67	67	—
6.	N-vinyl-N-ethylpropioneamide	70	70	70

Table 1.6 Calculated formation heats (*H*), dipole moments (*μ*), bond orders in C—N for *N*-vinylamides [48]. Reproduced by permission of Nauka

No.	Amide	Molecular mechanics		Quantum chemistry (PM3)			
		$-H$ (kJ/mol)	μ (D)	$-H$ (kJ/mol)	μ (D)	P	C^{cis}/C^{trans} (%)
1.	VF						
	cis-	290.0	3.65	93.47	2.74	1.092	99
	trans-	299.2	3.53	78.07	3.33	1.095	1
2	VAA						
	cis-	251.2	3.83	28.71	2.71	1.090	94
	trans-	253.4	3.72	27.02	3.50	1.069	6
3.	VMF						
	cis-	301.4	3.73	109.1	2.67	1.062	56
	trans-	309.6	3.37	108.1	2.74	1.062	44
4.	VMA						
	cis-	269.5	3.96	134.8	2.97	1.034	69
	trans-	274.7	3.51	113.6	3.35	1.042	31
5.	IPF						
	cis-	261.6	3.45	232.2	3.53	1.085	100
	trans-	248.2	3.84	215.9	3.16	1.124	0

Notes: abbrevations are as Table 1.3; IPF—*N*-isopropylformamide.

For an estimation of the conformational composition of *N*-vinylamides the fact was used that the highly intensive peak of the signal from the CH_3— group is upfield when it is located in close proximity to a C=O group, namely in the *cis*-position [49, 50]. A well-known example of such an arrangement is *N*-methyl-*N*-phenylformamide in which the intensive peak of the carbon atom signal belongs to the *cis*-conformer [50, 51].

trans-conformer (4%)   *cis*-conformer (96%)

Hence, *N*-vinylamides with alkyl substituent at the nitrogen atom are principally in *cis*-form and *N*-vinylamides with a proton at the same atom are principally in *trans*-form (Table 1.5).

Change of temperature in the range from 25 °C to 60 °C and the nature of the solvent has almost no effect on the conformer contents in *N*-vinylamides [47].

1.3 THEORETICAL CONFORMATIONAL ANALYSIS AND MONOMER ASSOCIATION OF *N*-VINYLAMIDES OF ALIPHATIC CARBOXYLIC ACIDS

Theoretical analysis of *N*-vinylamide conformations was realized by methods of molecular mechanics (PC Model Software) and quantum chemistry (PM 3 under the AMPAC Software) [48].

Experimental [52] and calculating (in brackets) [48] geometrical characteristics of amide group in *N*-vinylacetamide are represented in Scheme 1.1.

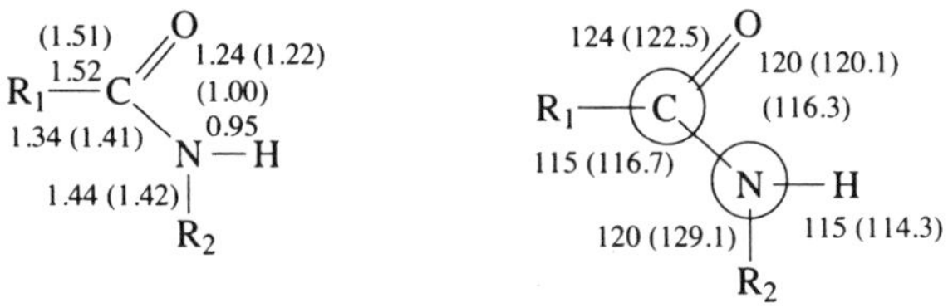

Scheme 1.1

It is seen (Scheme 1.1), that calculated geometry of a *N*-vinylacetamide molecule (left, bond length, right, valent angle) is close to the experimental one apart from bond length of C—N. The C—N bond is slightly lengthened which affects its order, and as a result, the barrier for internal rotation. The order (P) of a C—N bond in *N*-vinylamides (Table 1.6) varies in the range from 1.03 to 1.09, but achieves 1.12 in *N*-isopropylformamide. Accordingly, the calculated energy of

Table 1.7   Calculated formation heat energy (*H*), dipole moments (*μ*) of associates and heat of association of monomers (−Δ*H*) [48]. Reproduced by permission of Nauka

No.	Monomer	*H* (kJ/mol)	*μ* (D)	Δ*H* (kJ/mol)
1.	*N*-vinylacetamide (VAA) associate (VAA)$_2$			
	cis-cis	−249.8	3.10	−9.62
	cis-trans	−242.5	7.54	−9.12
	trans-trans	−239.8	7.89	−13.68
	associate (VAA)$_3$			
	cis-cis-cis	−383.6	1.76	−23.2
	trans-trans-trans	−374.8	12.74	−35.7
2.	*N*-vinyl-*N*-methylacetamide (VMA) associate (VMA)$_2$			
	cis-cis	−248.9	0.66	−15.8
	cis-trans	—	1.34	−14.7
	trans-trans	−260.9	1.76	−33.7
3.	*N*-vinyl-*N*-methylformamide (VMF) associate (VMF)$_2$			
	cis-cis	—	0.66	−19.5
	cis-trans	—	1.7	−24.6
	trans-trans	—	2.7	−8.45

the rotation barrier around the C—N bond differ: 97.9 kJ/mol for the latter in contrast 56.5 kJ/mol for the former. These values correspond to the data given in ref. [53].

Quantum-chemical calculations predict coplanarity for atoms of the amide group which is in accordance with röntgen ray analysis [53] and data of electron diffraction measurements [54].

Both calculating methods (molecular-mechanic and quantum-chemistry) prefer the *cis*-conformer for all compounds studied, namely those containing the dominating *cis*-form in the gas phase.

$$CH_2 = CH$$
$$N - C$$
$$R_1 \qquad R_2 \qquad O$$

cis-conformer

However, the calculated results (Table 1.6) are in conflict with ^{1}H and ^{13}C NMR data (Table 1.5) for *N*-vinylformamide and *N*-vinylacetamide. In fact, their *trans*-form contents are considerably reduced (1% and 6% calculated against 70% and 100% experimental, respectively), whereas the close conformity between these parameters is displayed for *N*-vinyl-*N*-methylformamide and *N*-vinyl-*N*-methylacetamide (see Tables 1.5 and 1.6).

The reason of such divergence between calculated (for gas medium) and experimental (for solvent polar medium) data in the case of *N*-vinylformamide and *N*-vinylacetamide is the fact that the association of monomer molecules in polar medium changes a conformational state of monomers to the more polar *trans*-conformer side [48]. The association of molecules is caused by both hydrogen bonds and dipole–dipole interactions.

The data of quantum-chemical calculations concerning hydrogen-connected associates (dimer and trimer) of *N*-vinylacetamide molecules containing a movable hydrogen atom, and dipole–dipole associates for molecules of *N*-vinyl-*N*-methylacetamide or *N*-vinyl-*N*-methylformamide incapable of hydrogen bond formation, is shown in Table 1.7.

Trans-trans dimer from *N*-vinylacetamide is formed with moderately strong hydrogen bonds ($-\Delta H = 13.4$ kJ/mol), having bond length 2.85 Å and a significant dipole moment (7.89 D) (Table 1.7), whereas *cis-cis* dimer of the same monomer is an associate having low enthalpy and low dipole. It is essential that a high dipole moment (12.8 D) arises at formation of a trimer (Fig. 1.2) from *trans-trans-trans*-conformers of *N*-vinylacetamide (Table 1.7). Calculations show that it is impossible to form normal hydrogen bonds in *cis-cis-cis* (VAA) trimer due to sterical difficulties.

N,N-substituted amides form only dipole–dipole associates. The calculations predict the existence of a very stable *trans-trans* VMA associate (Fig. 1.3) with a low dipole moment of 1.76 D because of the antiparallel mutual arrangement

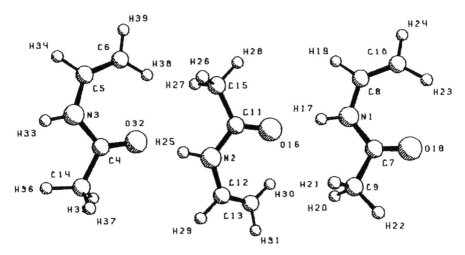

Fig. 1.2. Spatial structure of trimeric *N*-vinylacetamide (the *trans-trans-trans* form) with hydrogen bonds [48]. Reproduced by permission of Nauka.

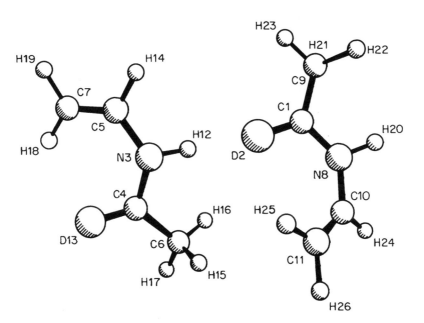

Fig. 1.3. Spatial structure of the dipole–dipole associate of *N*-vinyl-*N*-methylacetamide (the *trans-trans* form) [48].

of the C=O bonds. The dimeric *cis-cis* associate (Table 1.7) has an almost zero dipole moment and, although its heat of formation is lower than that of the *trans-trans* form, this percentage in the liquid can be fairly high.

Table 1.7 shows that linear associates of two monomers, containing a mobile proton at nitrogen atom in a *trans*-conformation, have the greatest dipole moments and association heat energy in comparison with associates in *cis*-conformation. Therefore, the conformational structure in a liquid in comparison with a gas phase is redistributed for the benefit of the dominant *trans*-conformation due to the formation of hydrogen bonds between molecules. That is why *N*-vinylacetamide in the liquid state is a *trans*-conformer and *N*-vinylformamide contains 70% of the same form.

From these data the important conclusion follows that extremely large dipole moments of linear associates created by *trans*-conformers of *N*-alkylamides cause abnormally large permittivities (ε), as found in experiments, as against low polar ones from *N,N*-alkylamides for which it is impossible to expect large ε values. In fact, high permittivity (ε) for amides involving the $-\overset{\overset{\text{O}}{\|}}{\text{C}}-\text{NH}-$ group (up to $\varepsilon = 180$) [55] against *N,N*-dialkylamides that have, as a rule, $\varepsilon = 36$–38, is known. The close values of ε (37.8 and 36.7) and dipole moments (3.81 and 3.86) are characteristic for $(\text{CH}_3)_2\text{N}-\text{C}\overset{\diagup\text{O}}{\diagdown\text{CH}_3}$ and $(\text{CH}_3)_2\text{N}-\text{C}\overset{\diagup\text{O}}{\diagdown\text{H}}$. On the contrary, monosubstituted amides, for example, $\text{CH}_3-\overset{\overset{}{\underset{\underset{\text{H}}{|}}{\text{N}}}}{}-\text{C}\overset{\diagup\text{O}}{\diagdown\text{CH}_3}$ (*trans*-form) and $\text{CH}_3-\overset{\overset{}{\underset{\underset{\text{H}}{|}}{\text{N}}}}{}-\text{C}\overset{\diagup\text{O}}{\diagdown\text{H}}$ (*trans*-form) with dipole moments of 3.73 and 3.83, respectively, are high polar liquids with $\varepsilon = 179$ and 183 [55].

Hence, the unusual dielectric properties of *N*-monosubstituted amides are stipulated by interactions between molecules being in *trans*-conformation that form long linear chains (associates) by means of hydrogen bonds.

It is evident that the knowledge of conformational states of non-cyclic *N*-vinylamides established by ^{13}C and ^{1}H NMR and theoretical calculations will be of great interest in future when reactivity of the monomers will be studied in various chemical reactions, for example, polymerization or copolymerization.

1.4 CONFORMATION OF RINGS IN *N*-VINYLLACTAMS

N-VINYLPYRROLIDONE

VP monomer structure contains a side ring consisting of five atoms and a plane amide bond $-\overset{\overset{\text{O}}{\|}}{\underset{}{\text{N}}}-\text{C}-$. For a reduction of stress in a ring caused by screened conformations of three methylene groups, a pyrrolydone ring similar to a substituted cyclopentane [56] could receive a non-plane distorted form. However, the rigid amide bond in the ring hinders oscillations of atoms upwards and downwards, as against cyclopentane. Quantum-chemical calculations display the planar arrangement for three atoms in $-CH_2-\overset{|}{N}-C=O$. The other carbon atoms are partially developed from a plane (partial propeller-type) with an insignificant deviation.

N-VINYLCAPROLACTAM (VCL)

As VCL ring is a seven-membered ring, it is not in a plane conformation.

As is well known [56], a seven-membered cycloheptane ring possesses various conformers a few types: 'chair', 'bath', 'twist-bath' and others. However, cycloheptane having a rigid double bond and 4-cyclohepten-4-on-1 having a double bond and the C=O group, both have no pseudo-rotation (transitions 'chair'–'bath'); as a result, the number of their conformers is reduced. In fact, the formation of the 'bath' conformer was found to be unfavourable due to repulsion of carbon atoms arranged on the 'nose' of the 'bath' and of atoms of the double bond. The most favourable conformation of cycloheptene is a 'chair' as for 4-cyclohepten-4-on-1 [56].

As the seven-membered VCL ring contains a rigid amide bond (partially double one $-N-\overset{\overset{\displaystyle O}{||}}{C}-$) and C=O group, the 'chair' conformation of VCL is most probable. This statement is also supported by ring conformation of caprolactam without a N-substituent.

Winkler and Dunitz [57] made an X-ray analysis of caprolactam displaying a 'chair' type of conformation. In crystal lattice the compound has a monoclinic space group of symmetry C $2/c$ with the following performance: $a = 19.28$ Å (2), $b = 7.78$ Å (1), $c = 9.57$ Å (1), $\beta = 112.39$ (10)°, $Z = 4$ (number of molecules in unit cell). The crystal lattice is based on a centre-symmetrical pair of molecules interacting through a hydrogen bond. The length of the $-NH\cdots O=C-$ bond is 2.90 Å, while the shortest possible distance between carbon methylene atoms of different rings is 3.65 Å.

As the caprolactam ring of VCL contains a vinyl group, one could expect a change of the performance of unit cell. In addition, the knowledge of conformational state of VCL molecules is required for the establishment of the relationship between monomer structure and reactivity in polymerization and copolymerization reactions, as well as of the physicochemical properties of polymers in water.

In order to determine the structural VCL performance two approaches may be used. The first is a theoretical one, using methods of molecular mechanics (PC

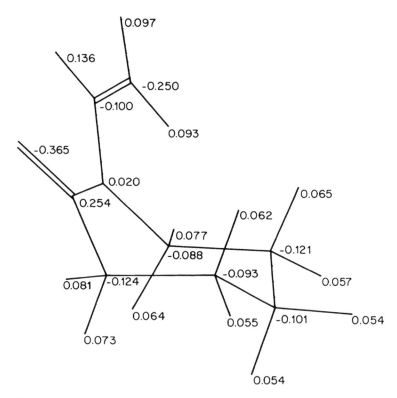

Fig. 1.4. Conformational structure of VCL computed by theoretical methods.

Model under MMX software) and quantum chemistry (PC-3 under the AMPAC software). The second is X-ray analysis of VCL crystals. It is of great interest to compare the data obtained by two approaches as it could give additional information relating to VCL properties as a monomer for radical polymerization.

Theoretical calculations allowed one to determine VCL conformation (Fig. 1.4) and its structural performance (bond length, valent and torsion angles,

Table 1.8 Conformational analysis of ring in VCL

Conformer	'Chair-1'	'Chair-2'	'Chair-3'	'Chair-4'	'Chair-5'
ΔH(kJ/mol)	0	5.02	5.14	5.39	12.9

Conformer	'Bath-1'	'Bath-2'	'Bath-3'	'Bath-4'
ΔH(kJ/mol)	19.5	24.9	25.9	27.32

charge distribution and so on) (Tables 1.9, 1.10 and 1.11). The calculations confirmed that the 'chair' conformation of VCL is the most stable whereas the 'bath' one is energetically unfavourable. The values of ΔH (difference of formation energy for various conformers) are represented in Table 1.8, confirming preferential ('chair-1') conformation of VCL.

The structure of 'chair-1' conformation of VCL with atoms being designated by number is represented schematically as follows:

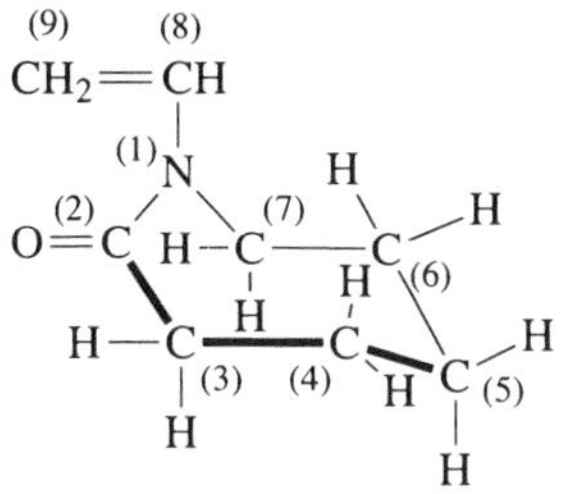

Scheme 1.2

Numerical values of atoms are given to ease the presentation of theoretical and X-ray data in Tables 1.9–1.11.

The same conformation was determined experimentally by X-ray analysis of VCL crystals by a 'Syntex-P2$_1$' diffractometer with CuK$_\alpha$-radiation ($\lambda = 1.5418$ Å). VCL structure (Fig. 1.5) is solved by direct method and refined to the R-factor being 0.062 [56].

Table 1.9 Length of bonds in VCL determined by X-ray analysis and quantum-chemical calculations (QCC) [58] and in caprolactam from X-ray analysis [57]

	Length of bonds (Å)		
Bonds	Caprolactam by X-ray	VCL by X-ray	VCL by QCC
C(2)—O	1.242	1.213	1.221
N—C(2)	1.327	1.381	1.434
C(2)—C(3)	1.501	1.519	1.521
C(3)—C(4)	1.519	1.536	1.522
C(4)—C(5)	1.522	1.519	1.519
C(5)—C(6)	1.525	1.533	1.519
C(6)—C(7)	1.509	1.526	1.526
C(7)—N	1.470	1.475	1.486
N—C(8)		1.407	1.4421
C(8)—C(9)		1.319	1.331
C—H	0.99–1.1	0.96–1.12	1.0–1.1

Table 1.10 Valent angles in caprolactam (CL) determined by X-ray analysis [57] and in *N*-vinylcaprolactam (VCL) from X-ray analysis and quantum-chemical calculations (QCC) [58]

	Valent angles, ± 0.2–0.3°		
Bonds	CL (X-ray analysis)	VCL (X-ray analysis)	VCL (QCC)
O—C(2)—N	120.9	122.0	117.5
O—C(2)—C(3)	120.6	121.6	122.5
N—C(2)—C(3)	118.5	122.0	120.0
C(2)—C(3)—C(4)	113.6	113.7	112.8
C(3)—C(4)—C(5)	113.9	114.0	114.0
C(4)—C(5)—C(6)	114.8	114.0	113.0
C(5)—C(6)—C(7)	113.9	113.8	114.1
C(6)—C(7)—N	113.7	111.9	112.8
C(7)—N—C(2)	125.5	116.4	114.6
C(8)—N—C(2)		117.6	119.2

Table 1.11 Torsion angles in a caprolactam (CL) molecule [57] and in a *N*-vinylcaprolactam (VCL) molecule determined by X-ray analysis [58]

	Torsion angles	
Bonds	CL	VCL, ± 0.3–0.4°
C(7)—N—C(2)—C(3)	– 4.2	– 4.37
N—C(2)—C(3)—C(4)	– 63.7	– 67.36
C(2)—C(3)—C(4)—C(5)	81.9	86.51
C(3)—C(4)—C(5)—C(6)	– 63.5	– 62.06
C(4)—C(5)—C(6)—C(7)	60.7	58.11
C(5)—C(6)—C(7)—N	– 77.0	– 80.57
C(6)—C(7)—N—C(2)	67.8	72.30
C(7)—N—C(2)—O	176.6	178.55
C(8)—N—C(2)—C(3)		172.38
C(8)—N—C(2)—O		– 4.71

VCL molecules form crystals of monoclinic structure P1 with performance: $a = 8.170$ Å (4), $b = 8.094$ Å (4), $c = 6.799$ Å (4); $\alpha = 99.92$ (1)°, $\beta = 88.89$ (1)°, $\gamma = 115.30$ (1)°, $Z = 2$, $V = 339.66$ Å^3. Experimental and theoretical performance of the VCL molecule is summarized in Tables 1.9–1.11.

Let us consider the packing of VCL molecules in a crystal (Fig. 1.6). A unit cell consists of two VCL molecules placed on top of one another with their 'backs' in opposite directions.

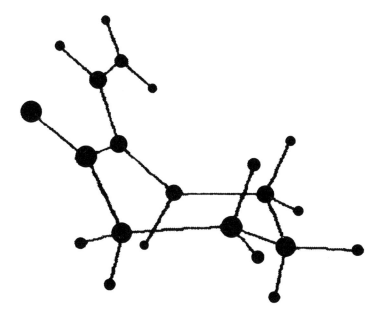

Fig. 1.5. The structure of VCL molecule found experimentally by X-ray analysis of monomer crystals [58].

Three planes are observed in the VCL molecule (Fig. 1.5). The first (a 'back') is composed of vinyl and amide groups involving C(3) and C(7), C(9), C(8), N, C(2), O, C(3) and C(7) with a plane deviation of + 0.088, + 0.026, −0.037, −0.014, −0.027, +0.098 and −0.053 Å respectively.

The second plane involves C(7), C(3), C(4) and C(6) with plane deviations of +0.017, −0.017, −0.028 and +0.019 Å respectively. Three carbon atoms, namely C(6), C(5) and C(4) with plane deviations 0.000, 0.000 and 0.000 are organized in the third plane.

Thus, both theoretical and experimental methods display 'chair' conformation in VCL monomer. It is essential that the performance data (the length of valent bonds, valent and torsion angles and so on) obtained by X-ray analysis and quantum-chemical calculations coincide. Some difference in length of valent bond found by the two methods is observed only for the amide group. Close values of structural parameters (experimental and theoretical) allow one to apply a theoretical approach in charge distribution estimation on atoms of VCL and other *N*-vinylamide monomers.

Charge values on VCL atoms and chemical shifts of carbon atoms (^{13}C) and protons (^{1}H) obtained from NMR spectra of VCL in CDCl$_3$ are represented in Table 1.12.

Table 1.12 Chemical shifts of carbon atoms in the ^{13}C NMR spectrum and protons in the ^{1}H NMR spectrum of VCL

Atoms	Chemical shifts	
	^{13}C(ppm)	^{1}H(ppm)
CH$_2$=	89.06	4.00
—CH=	128.60	6.96
C=O	170.6	
CH$_2$(3)	38.8	2.12
CH$_2$(4)	23.9	1.33
CH$_2$(5)	20.15	1.26
CH$_2$(6)	26.00	1.39
CH$_2$(7)	40.5	3.2

Notes: 'Bruker-AMX-400', CDCl$_3$.

1.5 ELECTRONIC STATE OF MONOMERS

The double bond in a *N*-vinylamide molecule is in conjunction with carbonyl through a nitrogen heteroatom. Therefore, changes of solvent nature are reflected in an electronic state of double bond. It was found that in the IR spectra of VP in D$_2$O solution, one band of stretching vibrations of C=O group gradually shifts from 1695 to 1650 cm^{-1} as the solution is diluted, whereas the other band at 1630 cm^{-1} (stretching vibrations of CH=CH—) is almost unaffected by dilution. Hence, reduction of VP concentration in aqueous solution results in an appreciable shift of an absorption band, associated with the C=O group, to lower frequencies, leaving the bond associated with double bond unchanged. The shift to lower frequencies predominantly takes place in a concentration range extending from pure VP to its 10% aqueous solution. Further decrease of VP concentration in D$_2$O solution produces almost no effect on the position of C=O absorption band. A similar variation pattern for the position of the C=O band was established also for *N*-ethylpyrrolidone [60], a saturated analogue of VP.

The absence in the IR spectra of VP of some expected bands related to the stretching vibrations of the C=O group is caused by the multiple hydrogen bonds characterized by different bond energies, in which the C=O group of pyrrolidone ring and water molecule participate.

Table 1.13 Position of the band due to stretching vibrations of C=O for some N-vinyl-amides in different solvents [20]. Reproduced by permission of Nauka

	$\nu_{C=O}$ cm^{-1} ± 2		
Solvent	N-vinylpyrrolidone (VP)[a]	N-vinylcaprolactam (VCL)[a]	N-vinyl-N-methylacetamide (VMA)[a]
Pure compound	1695	1670	1650
CDCl$_3$	1680	1665	—
C$_2$D$_5$OD	1675	1647	—
D$_2$O	1650	1620	1610

[a]At 10% monomer solutions; [VCL] in water is 1%.

The decrease of $\nu_{C=O}$ with dilution (VP content reduces from 50% to 10%) is attributed to the formation of a hydration shell in which the proton- (deutron-) donating ability of water molecules involved in the C=O···D—O—D hydrogen bonds is enhanced due to the formation of hydrogen bonds between water molecules within this shell. Spin-lattice relaxation of H$_2$O protons in a mixture with VP shows that hydration shell of VP molecules could accommodate not more than 58 molecules of water [61].

When water was replaced by organic protic solvents, the decrease in $\nu_{C=O}$ of N-vinylamides was not so significant (Table 1.13) [20].

An essential change of electronic density on carbon atoms of N-vinylamides being in protic solvents finds reflection in ^{13}C NMR spectra depending on mixture composition (monomer + water) [59]. NMR studies of VP in aqueous solution revealed a remarkable shift of signal positions of carbons atoms in C=O and =CH$_2$ groups downfield at reduction of VP concentration in water (Fig. 1.7), while variation in chemical shifts of carbon atom in —C*H$_2$—C=O and —CH$_2$—C*H$_2$—CH$_2$— groups remains insignificant.

It is interesting that a reduction of VP contents in solution from 100 to 10 wt% is accompanied by a cymbate shift of the signals of carbon atom in C=O and CH$_2$= downfield (Fig. 1.7). Further VP concentration reduction does not affect the chemical shift. The observed shift of the signal of carbon atom in a CH$_2$= group of double bonds to downfield when VP is diluted by water indicates a certain decrease of electron density on the carbon atom, whereas a small shift of the signal of the carbon atom in a —CH= group upfield is caused by an insignificant increase of electron density.

The data on ^{13}C NMR and IR spectra of VP in aqueous solution testifies that the expansion of the solvation shell (H$_2$O or alcohol) via hydrogen bonding increases the polarity of the double bond, owing to its conjugation with the C=O group via the nitrogen atom.

Changes of a similar character in IR and ^{13}C NMR spectra were found for VCL [62] and N-vinyl-N-methylacetamide [63] that assumes the general charac-

Table 1.14 Electron density distribution on atoms of *N*-vinylamides[a]

Monomer	$CH_2=$	$=CH-$	N	$\overset{a}{C}=O$	$C=\overset{a}{O}$	$\overset{O}{\underset{C-H}{\parallel}}{}^a$	$\overset{O}{\underset{C-CH_3}{\parallel}}{}^a$
$CH_2=CH \quad O$ $\quad\underset{NH-C-H}{\mid}\quad\overset{\parallel}{}$ *N*-vinylformamide	-0.243	-0.165	$+0.012$	$+0.227$	-0.372	$+0.098$	
$CH_2=CH \; O$ $\underset{H}{\overset{\mid}{N}}-\overset{\parallel}{C}-CH_3$ *N*-vinylacetamide	-0.245	-0.111	$+0.04$	$+0.239$	-0.386		-0.13
$CH_2=CH$ $\underset{CH_3}{\overset{\mid}{N}}-C\overset{\nearrow O}{\underset{\searrow H}{}}$ *N*-vinyl-*N*-methylformamide	-0.298	-0.102	-0.01	$+0.240$	-0.365	-0.093	
N-vinyl-*N*-methylacetamide	-0.247	-0.102	-0.01	$+0.257$	-0.379		-0.14

[a] PM-3 on AMPAC software.

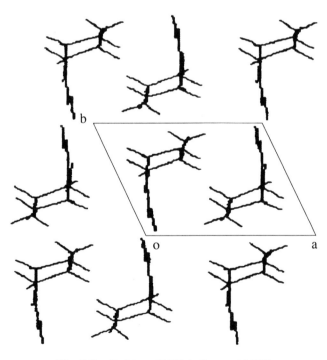

Fig. 1.6. Packing of VCL in the crystal [58].

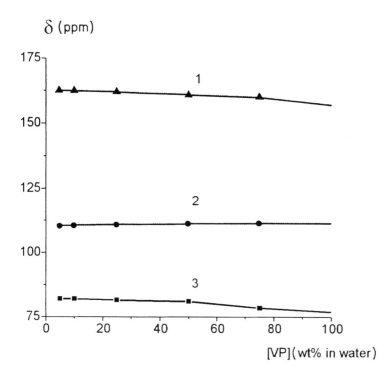

Fig. 1.7. Dependence of chemical shift of ^{13}C in $C{=}O$ (1), $-CH{=}$ (2) and $CH_2{=}$ (3) groups of VP on its concentration in water. Reference point is chemical shift (δ) in $-CH_2- C^*H_2-CH_2-$ group [59]. Reproduced by permission of Nauka.

ter of influence of protic solvents on the electronic state of the double bond in the N-vinylamides investigated.

Thus, as exemplified by the NMR data concerning VP, VCL and other N-vinylamides, the electronic state of the double bond in monomers may considerably vary with variations of solvent nature and concentration, the related effects being most pronounced in protic solvents.

In fact, these monomers are characterized by significant partially negative charges on the oxygen atom in $C{=}O$ and the carbon atom in $CH_2{=}$ of the vinyl group (Table 1.14). A large number of water molecules in the hydration shell (58) surrounding a VP molecule may be caused by large charges on $C{=}O$ which participate in the formation of a network of hydrogen bonds that surrounds the monomer. The marked negative charge on $CH_2{=}$ of the double bond and the conjunction with a carbonyl will define reactivity of the monomer in polymerization and copolymerization reactions and high sensitivity of monomer reactivity to solvent nature.

Table 1.15 Charge distribution on VP molecule atoms at different number of water molecules surrounding a monomer

Number of water Molecules	$\overset{a}{C}H_2=$	$-\overset{a}{C}H=$	$\overset{a}{C}-O$	$C-O^a$
	Charges on atoms			
0	-0.237	-0.097	$+0.259$	-0.340
1	-0.232	-0.100	$+0.272$	-0.352
5	-0.222	-0.105	$+0.278$	-0.369
10	-0.212	-0.12	$+0.291$	-0.391

[a] PM-3 method on the AMPAC software.

The distribution of charges on all atoms of VP is submitted in Scheme 1.3.

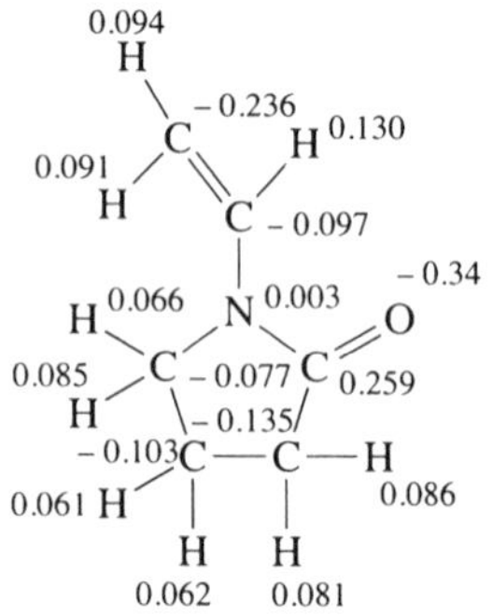

Scheme 1.3

Quantum-chemical calculations show that a change at hydration surrounding of VP molecule considerably affects the charge distribution on carbon and oxygen atoms of *N*-vinylamide monomers. Partially negative charges on an oxygen atom of C=O increases, and on the carbon atom of $CH_2=$ of the double bond decreases, with increasing the number of water molecules (Table 1.15).

It is important to note that the increase of the number of water molecules results in the reduction of negative charge both on carbon atoms in $CH_2=$ and in C=O, and its enhancement on the carbon atom of the methylene group, correlating with signal shifts of ^{13}C NMR spectra of VP (Fig. 1.7) with decreasing monomer concentration in water.

The high negative charge on the carbon atom in $CH_2=$ also dictates the significant reactivity of the double bond when interaction with H^+ takes place. The quantum-chemical calculations show that the affinity energy of the $CH_2=$ group to H^+ ($E = 828.8$ kJ/mol) markedly exceeds that of the oxygen atom in C=O (797 kJ/mol) and that of the =CH— group (591.6 kJ/mol). As a result, a carbocation is formed that participates in various reactions (1.32, 1.33).

$$\overset{-\delta}{CH_2}=CH \qquad \qquad CH_2-\overset{+}{CH}$$

$$\left[\underset{N}{}\right]\underset{C}{\overset{-\delta}{=}}O + H^+ \longrightarrow \left[\underset{N}{}\right]\underset{C}{=}O \qquad (1.32)$$

Depending on acid concentration, VP in aqueous solutions can add a water molecule forming N-(α-hydroxyethyl)pyrrolidone [62–64] or split to acetaldehyde and pyrrolidone [65].

$$CH_2=CH \qquad \qquad CH_3-CH-OH$$

$$\left[\underset{N}{}\right]C=O + H_2O \xrightarrow{H^+} \left[\underset{N}{}\right]C=O \qquad CH_3-C\overset{O}{\underset{H}{\diagdown}} + \left[\underset{NH}{}\right]C=O$$

$$\text{or}$$

$$(1.33)$$

Therefore, the polymerization reaction of N-vinylamides is carried out only in neutral or alkaline medium. Realization of the reaction in acidic media can considerably distort experimental data [66].

2

Radical Polymerization

2.1 PREVIOUS STUDY OF *N*-VINYLPYRROLIDONE POLYMERIZATION

For the first time, water-soluble polymer products of various molecular weights (MWs) were prepared from *N*-vinylpyrrolidone by Fickentscher and Herrle in 1939 [67, 68]. They developed a method of polymerization of this monomer in water in the presence of hydrogen peroxide and ammonia.

This method then was fixed as a basis of technological process of poly-*N*-vinylpyrrolidone (PVP) preparation for medicine [67–69]. PVP with MW $40–50 \times 10^3$ was used for manufacturing medicinal preparations, having a property of plasma substitute (3.5% water–salt solution of PVP) [70]. The very earliest use of PVP in medicine was during the Second World War when the solution mentioned above was infused into patients as a synthetic blood plasma volume expander. About 500 000 soldiers of the German Army during the Second World War received an injection of this preparation. The high biological activity of this polymer, absence of toxicity, stability to thermal processing in water and a number of other properties promoted its wide application in medical practice.

It is of interest to note the peculiarities of the VP polymerization process in a water solution in the presence of H_2O_2 and NH_3, found by its developers (Fickentscher and Herrle) [11, 67–69].

1. Polymerization should be carried out in buffer, neutral or poorly alkaline solutions to avoid hydrolysis of the monomer with the formation of acetaldehyde, and in subsequent acetic acid.
2. The reaction rate depends on the concentration of added ammonia or amine. The reaction does not proceed at introduction of NaOH or KOH. The constant K (Fickentscher's constant), characterizing MW of a polymer, does not, in practice, depend on ammonia concentration.
3. An enhancement of temperature slightly increases the polymerization rate, but affects MW only insignificantly.

4. The reaction rate is increased proportionally to the square root of the peroxide concentration.
5. It increases with growth of initial monomer concentration up to the maximum value at a concentration equal to 35%, remains constant with the concentration ranging from 35% to 60%, and decreases again at further increase of monomer concentration. The values of MW of polymer obtained do not, in practice, depend on the latter.
6. This reaction is inhibited by oxygen.

The description of VP polymerization process in industry is reported in [69]. The polymerization of VP was carried out at 80 °C. In the beginning of the process half of a solution volume was used, and the rest was added gradually during two to three hours. As an activator, hydrogen peroxide was applied with 0.05–2% concentration in the reaction mixture [11, 67].

However, such a procedure resulted in widening of MW distribution of PVP obtained because of partial decomposition of an activator during a prolonged reaction time. In this case, the proportion of fractions of high MW in the polymer increased [70].

Hereafter, this approach was improved to prepare PVP with various MWS and narrow MW distribution. During the last 30–40 years the improved process was basic in fabricating PVP by firms in Germany (BASF), USA (GAF, later ISP) and Russia.

The isolation of polymers from water solutions as a powder is achieved by spraying in a dryer. Because of the presence of impurities (residual monomer, hydrolysis products of the monomer and other impurities) one cannot use the polymer as a component of a blood-plasma substitute. Therefore, dried powder of the polymer, as reported in [69], is exposed to monomer extraction by organic solvent, for example, methylene chloride. Removal of monomer from the polymer is conducted also by an extraction of a polymer solution, received after the polymerization, by organic solvent [2].

In 1970s and 1980s a series of patents devoted to preparation of higher quality PVP was issued [71–76]. These patents were aimed at the development of PVP technology free from shortcomings of the traditional process of medical PVP synthesis in the presence of H_2O_2, including the low polymer yield, large volume of organic solvent used to extract the monomer, the presence of impurities of unknown structure and wide MW distribution.

2.2 INITIATORS AND REACTIONS WITH ITS PARTICIPATION

SYSTEM OF $H_2O_2 + NH_3$

For the first time, the initiation of radical VP polymerization was realized in the system of H_2O_2 and NH_3 [2, 11, 67–69]. The amount of H_2O_2 added ranging from 0.1% to 3% (30% H_2O_2) relative to monomer concentration varies the polymerization rate and MW of polymers obtained. In such a way PVP with $\bar{M}_w$ ranging from 10×10^3 to 100×10^3 was synthesized.

For a long time the initiation mechanism of this traditional process remained unknown; this complicated the understanding of the radical polymerization peculiarities mentioned above [2, 69, 76]. The comprehension of the role of hydrogen peroxide in initiation of the polymerization reaction and in MW regulation of PVP received, and of the effects of solvent nature and impurities of metal ions on the reaction rate is a necessary precondition in the development of the process of high-grade PVP production.

The use of pure reagents (gaseous ammonia, water with iron ion contents less than 10^{-8} mol/l and monomer containing no amino impurities) results in the complete cessation of polymerization, indicating reduction–oxidation type of initiation reactions. The application of NaOH or KOH instead of ammonia to keep the pH at 9.0 also terminates the process, although the system contains trace amounts of iron impurities [67, 76]. Most favourable content of heavy metal ions, necessary for initiation of the polymerization process, lays in a range from 10^{-5} to 10^{-6} wt%. In this case, an addition of ammonia can be ruled out. This reaction originates already in the presence of NaOH and KOH. An application of $CuCl_2$, $FeCl_3$ or $CoCl_2$ as ions of heavy metals was proposed [77].

Let us consider the VP initiation reaction in water in the presence of H_2O_2 and NH_3. This reaction occurs due to a number of reduction–oxidation reactions (2.2 and 2.4) between H_2O_2 and external iron (ferric hydroxide) involved in a complex with NH_3 [78, 79]. In the presence of ammonia the formation of combined complexes of iron occurs, where atoms of nitrogen and oxygen act as ligant atoms (2.1).

$$
\begin{array}{c}
O \\
| \\
O-Fe^{3+} + NH_3 \\
| \\
O
\end{array}
\longrightarrow
\begin{array}{c}
O \\
| \\
O-Fe^{3+}----NH_3 \\
| \\
O
\end{array}
\qquad (2.1)
$$

$$H_2O_2 \longleftrightarrow H^+ + HO_2^- \qquad (2.2)$$

$$
\begin{array}{c}
O \\
| \\
O-Fe^{3+}----NH_3 + HO_2^- \\
| \\
O
\end{array}
\longrightarrow
\begin{array}{c}
O \\
| \\
O--Fe^{2+}----NH_3 + HO_2^{\bullet} \\
| \\
O
\end{array}
\quad (2.3)
$$

$$
\begin{array}{c}
O \\
| \\
O-Fe^{2+}----NH_3 + H_2O_2 \\
| \\
O
\end{array}
\longrightarrow
\begin{array}{c}
O \\
| \\
O-Fe^{3+}----NH_3 + HO^{\bullet} + HO^- \\
| \\
O
\end{array}
\quad (2.4)
$$

At heavy metal ion (Fe^{3+}, Cu^{2+}, Co^{2+}) concentration less than 10^{-6} wt%, the reaction rate is lowered because of a low concentration of $OH^{\bullet}$ radicals formed.

At high ion concentration ($>10^{-4}$ wt%), oxygen is formed, inhibiting the polymerization reaction.

$$\underset{\overset{|}{O}}{\overset{\overset{O}{|}}{O-Fe^{3+}}}----NH_3 + HO_2^{\bullet} \longrightarrow \underset{}{\overset{\overset{O}{|}}{O-Fe^{2+}}}----NH_3 + O_2 + H^+ \quad (2.5)$$

The $HO^{\bullet}$ radical (reaction 2.4) interacts with the double bond of VP, causing radical polymerization. However, it is necessary to take into account secondary reactions with participation of $HO^{\bullet}$.

The reaction of $HO^{\bullet}$ radicals with double bonds of VP (with the reaction rate constant (K) being $(7.0 \pm 0.6) \times 10^9$ l/mol s) is the most preferable in comparison with an abstraction reaction of the hydrogen atom from PVP ($K = 0.2 \times 10^9$ l/mol s) and N-methylpyrrolidone ($K = 2.6 \times 10^9$ l/mol s) [80].

$$HO-CH_2-\overset{\bullet}{C}H$$
$$\underset{\overset{|}{CH_2}-CH_2}{\overset{CH_2}{\diagdown}}\overset{N}{\underset{\diagup}{}}\overset{}{C=O}$$

I

The kinetic chain at VP polymerization in water with $H_2O_2 + NH_3$ commences with a chainlink of structure I [81, 82]. Thus, initiation of VP polymerization reaction in the mixture considered is caused by the presence of iron impurities in ferric hydroxide forms, which are always in water or in aqueous solutions in trace amounts. The introduction of NH_3 changes the reduction–oxidation potential of Fe^{3+} ions, promoting reaction of these ions with H_2O_2 in alkaline medium to form Fe^{2+}. It becomes obvious that the concentration of iron impurities is an important factor, affecting the polymerization reaction rate in the system VP + $H_2O + H_2O_2 + NH_3$.

In order to control the VP polymerization process in water with H_2O_2, it is recommended that ions of heavy metals should be included and to exclude ammonia as a substance, promoting the formation of toxic hydrazine [77].

ORGANIC PEROXIDES, UV- AND γ-IRRADIATION

To initiate the VP polymerization reaction, other peroxides and hydroperoxides, namely, dicumyl peroxide, di-*tert*-butyl peroxide, *tert*-butyl hydroperoxide, *tert*-butyl perbenzoate and others are used [71–73, 83]. This reaction in the presence of peroxides mentioned above is carried out at a temperature range from 140 to 300 °C depending on the initiator used.

In order to decrease the polymerization reaction temperature, the introduction of activators promoting peroxide decomposition in a reaction mixture, such as copper or magnesium acetate, copper acetylacetonate, copper stearate, cobalt naphthenate and other is proposed [83].

Azo-compounds have found wide application as initiators of *N*-vinylamide polymerization. So, azo-bis-isobutyronitrile is the most used initiator of polymerization reaction of VP [2, 84, 85], *N*-vinylpiperidone [85, 86], *N*-vinylcaprolactam [85, 87], *N*-vinyl-*N*-methylacetamide [88], *N*-vinylformamide [89, 90] and *N*-vinylacetamide [39].

It is necessary to note that benzoyl peroxide widely applied to initiate polymerization of large groups of vinyl compounds is an inefficient initiator of *N*-vinylamide polymerization in protic solvents. The polymerization reaction of such monomers as VP [2], VCL [87] or *N*-vinylacetamide [39] in alcohols in the presence of benzoyl peroxide is characterized by a low yield (10–30%). Such a pecularity of this reaction in the presence of the latter is caused by the formation of benzoic acid at thermal decomposition of this initiator: (reactions 2.6, 2.7) [91].

$$\text{Ph}-\overset{\overset{\displaystyle O}{\|}}{C}-O-O-\overset{\overset{\displaystyle O}{\|}}{C}-\text{Ph} \quad \longrightarrow \quad 2\,\text{Ph}-\overset{\overset{\displaystyle O}{\|}}{C}-O^{\bullet} \qquad (2.6)$$

$$\text{Ph}-\overset{\overset{\displaystyle O}{\|}}{C}-O^{\bullet} + C_2H_5OH \quad \longrightarrow \quad \text{Ph}-\overset{\overset{\displaystyle O}{\|}}{C}-OH + CH_3-\overset{\bullet}{C}H-OH \quad (2.7)$$

The occurrence of H$^+$ in an *N*-vinylamide system, as a result of benzoic acid formation, accelerates the rate of side reactions proceeding with participation of the carbocation (see Chapter 1) and bringing about the formation of *N*-(α-hydroxyethyl) amide derivatives, acetaldehyde or ethylidene-bis-amides in relation to the nature of the solvent.

Akashi and co-workers [39] discovered the decomposition of *N*-vinylacetamide in a polymerization system involving benzoyl peroxide and deuterium dimethylsulfoxide at 60 °C by NMR techniqque. In this case, the radical being formed as a result of chemical reaction (2.6) is likely to abstract proton (or deuteron) from a solvent molecule making a fresh start to the benzoic acid preparation.

Such an initiator as potassium persulphate also does not initiate *N*-vinylacetamide polymerization reaction in aqueous solutions as the monomer reacts with the former to give the hydrolyzates, as shown recently by NMR measurements [39].

It is well known that potassium persulphate decomposes thermally, giving sulphate radical ion which initiates vinyl polymerization [92].

Fergusson and Rajan [93] have reported VP hydrolysis reaction in the presence of potassium persulphate. In fact, the presence of the initiator greatly increases the rate of the VP hydrolysis at 74 °C. The evolution of acetaldehyde and pyrrolidone during the VP hydrolysis reaction in the presence of potassium persulphate was confirmed by analytical methods and IR-spectroscopic analysis The significant reactions probably include the following:

$$S_2O_8^{-2} \longrightarrow 2SO_4^{-} \qquad (2.8)$$

$$SO_4^{-} + H_2O \longrightarrow HSO_4^{-} + HO^{\cdot} \qquad (2.9)$$

$$HSO_4^{-} \longleftrightarrow H^+ + SO_4^{-2} \qquad (2.10)$$

The VP polymerization in the system does not occur to any measurable extent. Of the two competitive reactions (initiation of polymerization and hydrolysis), the latter is much faster. The hydrolysis of VP (2×10^{-2} mol/l in water) in the presence of $K_2S_2O_8$ (0.74×10^{-2} mol/l) was found to proceed completely at 74 °C for 20 min as the pH falls to 2.0 [93]. Hence, compounds of the structure mentioned above cannot be used as an initiator of free radical polymerization of N-vinylamides due to the high reactivity of double bond to proton with a carbocation formation (see Chapter 1).

Polymerization of VP [69, 80, 94], VCL [95, 96], N-vinyl-N-methylacetamide [88] and N-vinylacetamide [39] was performed under the action of γ-radiation (^{60}Co). A study of pulsing radiolysis process in VP aqueous solutions under the action of γ-radiation (^{60}Co) testifies that polymerization of the monomer is stipulated by hydrated electrons (e_{H_2O}) and HO$^{\cdot}$ radicals, formed under action of γ-radiation on water. However, the contribution to the initiation of polymerization reaction with the participation of hydrated electrons is insignificant and is only 5% [80]. The reaction of e_{H_2O} with the double bond of VP proceeds with a high rate constant ($\sim 1.6 \times 10^9$ l/mol s). The atoms of a ring virtually do not participate in interaction with hydrated electrons ($k_2 = 1.3 \times 10^7$ l/mol s). A HO$^{\cdot}$ radical in water also reacts with the double bond of VP. As a result, macromolecules contain terminal OH groups, with which chains begin, just as occurs in the system VP, H_2O, H_2O_2 and NH_3.

Radical polymerization of N-vinylamides can be initiated by UV-radiation on an aqueous solution involving a monomer and H_2O_2 [78]. The molecule of hydrogen peroxide decomposes under the action of UV light with the formation of two radicals which begin a chain.

2.3 SOLVENT EFFECTS

In monomers of N-vinylamide generation the vinyl group is joined to the C=O group through a nitrogen atom. Dipole moments of these compounds are high and lay in a range from 3.5 to 4.0 (Table. 1.7). Due to the large dipole moment and the presence of the C=O group, capable of hydrogen bond formation with molecules of a protic solvent, the electronic state of the vinyl group becomes solvent-dependent (see Chapter 1) which, in turn, is reflected in the reactivity of these monomers in the radical polymerization reaction.

N-VINYLPYRROLIDONE

The effects of the nature of the solvent on VP polymerization in a mixture with H_2O and other organic solvents were reported in refs [2, 11, 67, 70, 84].

In a mixture of VP with H_2O, H_2O and NH_3 the polymerization reaction rate grows with increasing monomer concentration in water up to 35%, remains constant with VP concentration ranging from 35% to 60% and decreases rapidly with concentration growth beyond 60% [67]. However, the interpretation of such dependence of the reaction rate on VP concentration in water is complicated due to secondary reduction–oxidation reactions. The initial polymerization reaction rate (V_o) depends on the initiation reaction rate which is quite sensitive to H_2O_2, NH_3 and iron impurity concentration [76, 78].

The effect of the medium on VP polymerization is also shown in other methods of initiation, for example, at γ- and UV-radiation and thermal decomposition of azo-bis-isobutyronitrile [76, 80, 95, 96].

Davis and Senoglas [80] found the effect of mixture compositions of VP with H_2O and other solvents on kinetic curves of VP polymerization in water, in ethylacetate and in benzene under γ-radiation (^{60}Co) action.

The initial rate of VP polymerization (V_o) in dilute water solutions (5, 10 and 25% VP concentration) is found to be four to five time greater than that in concentrated VP solution (50 and 75% concentration) or in pure VP. Also the V_o value in ethylacetate markedly exceeds that in benzene [80].

The role of the solvent in this reaction is shown by consideration of the dependence of the initial polymerization rate reaction (V_o) on monomer concentration in solvents of various kinds in the following equation (2.1):

$$k_{eff} = \frac{V_o}{[M_o]} = \frac{k_p}{k_t^{1/2}} \, (f k_{in} [J])^{1/2} \qquad \text{(eq. 2.1)}$$

where V_o is the initial rate of reaction, k_p, k_t and k_{in} the constants of reaction rate for chain propagation, chain termination and initiation, respectively, $[M_o]$ and $[J]$ the initial concentrations of a monomer and initiator and f the efficiency of initiation. In addition, initiation reaction rate (thermal decomposition of AIB*N*-azo-bis-isobutyronitrile (AIBN) [97] or photolytic decomposition of H_2O_2 [100]) remain constant for various compositions of mixtures investigated.

It is possible to detect regions of concentrations on a dependence curve of k_{eff} on [VP] (Fig. 2.1) in water, for which appreciable distinctions in reaction kinetics are observed.

Recently the VP polymerization kinetics in water at various monomer concentrations was studied by NMR method [Topchiev, D. A., Martynenko, A. I.,

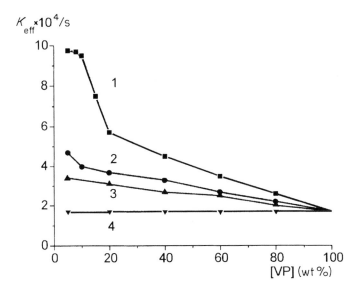

Fig. 2.1. Dependence $k_{eff} = V_o/[M_o]$ on VP contents in water (1), ethanol (2), 2-pyrrolidone (3) and N-methylpyrrolidone (4) [97]. Reproduced by permission of Nauka.

Kabanova, E. Yu and Timofeeva, L. M. (1997) *Vysokomol Soed.* **39A**: 1129–1139]. It was found that the enhancement of the k_{eff} value in going from 10 wt% aqueous solution of the monomer to the block is less significant (2-fold) than that of k_{eff} observed in ref. [94, 97, 100]. Dilatometric method has proved to be incorrect because the contraction coefficient is dependent on VP concentration in water.

The first region ranging from 1% to 10% is characterized by V_o proportional to the [VP], i.e. k_{eff} remains constant.

The second region ranging from 10% to 30% of [VP] is characterized by a sharp fall of the value of k_{eff}. With increasing monomer concentration from 30% up to 100% a progressive lowering of this constant occurs. The order of reaction on VP in a concentration range from 10% up to 100% becomes essentially less than one. The greatest value of k_{eff} occurs at [VP] < 10% and exceeds five times as much that in the block [92].

It should be noted that a similar change of the reaction rate is found in other methods initiation, namely at UV-radiation of H_2O_2 solutions (Fig. 2.2) [100]. Dependence of $k_{eff} = V_o[M_o]$ on [VP] at UV-radiation and that at thermal decomposition of AIBN are similar (Fig. 2.2).

In other protic solvents (ethanol and 2-pyrrolidone) the character of V_o dependence on [VP] is maintained, but the effect of solvent on reaction acceleration becomes essentially less than in the case of H_2O. So, for ethanol and 2-pyrrol-

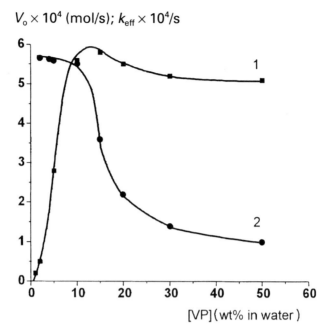

Fig. 2.2. Dependence of initial rate of VP polymerization (V_o) (1) and $k_{eff} = V_o/[M_o]$ (2) on monomer concentration in water (UV-light, $\lambda = 313$, [H_2O] = 0.1 mol/l) [100]. Reproduced by permission of Nauka.

idone the increase of k_{eff} totals 2.5 and 1.8 times respectively against k_{eff} for a pure monomer (Fig. 2.1). In *N*-methylpyrrolidone the value of k_{eff} remains constant in the whole concentration region, from 5% up to 100%.

The effect of protic solvent (water or butanol) on the rate of VP polymerization reaction under the action of γ-radiation is also reported in ref. [94].

N-VINYLCAPROLACTAM AND N-VINYL-N-METHYLACETAMIDE

In a study of kinetics of VCL radical polymerization in a mixture of solvents (water + *N*-methylpyrrolidone, as we can dissolve only 1% VCL in H_2O) at constant monomer concentration the growth of k_{eff} is found with increasing water content in a mixture [62]. So, at 25 wt% content of *N*-methylpyrrolidone in water k_{eff} becomes about six times greater than that in pure *N*-methylpyrrolidone.

For *N*-vinyl-*N*-methylacetamide the solvent effect on polymerization reaction acceleration is also found on going from a pure monomer to its water solutions [63].

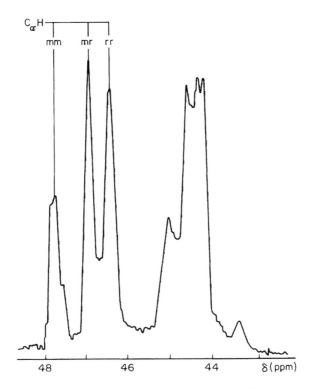

Fig. 2.3. Signals of chain —$C_\alpha H$— and ring —CH_2—N— in the ^{13}C NMR spectrum of PVP in D_2O, 50 °C, 7 mol%, spectrometer 'Bruker-WH-360' [106]. Reprinted from *Prog. Polym. Sci.* **18**, Kirsh, Yu. E. 519–542, © 1993, with kind permission from Elsevier Science Ltd, The Boulevard, Langford Lane, Kidlington OX5 1GB, UK.

Thus, the nature of protic solvents, in particular water, increases the rate of radical polymerization of a series of *N*-vinylamides (VP, VCL and VMA). Knowledge of this factor allows direct adjustment of the process of synthesis of water-soluble poly-*N*-vinylamides reducing the reaction time and increasing polymer yield.

The peculiarity of VP kinetic behaviour in radical polymerization reaction in solvents of various nature testifies that the reactivity of monomer grows at going from *N*-methylpyrrolidone, 2-pyrrolidone to ethanol and H_2O. In the same way [pK_a of 2-pyrrolidone (25) > pK_a of ethanol (18) > pK_a of H_2O (14) [99]) the acidity of solvents under investigation is increased.

The reaction rate depends on the molar ratio of monomer and water in a mixture. With increasing water contents from 0 to 3–5 molecules of H_2O on one molecule of VP, there is insignificant acceleration of polymerization reaction rate (1.2 times) (Fig. 2.3). The further growth of the number of H_2O molecules increases a hydrate shell to a close proximity of a monomer molecule with formation of stronger hydrogen bonds between $>N-\underset{|}{C}=O\cdots H-O-H$.

In a narrow range of VP concentration, from 20 to 10 wt%, when the number of H_2O molecules rises from 24 to 58, a sharp jump of the effective reaction rate constant is observed. If the number of water molecules exceeds 58 at one monomer molecule, k_{eff} does not change. Such molar ratio (58:1), established from the kinetic data, corresponds to the greatest number of H_2O molecules in hydrate shell surrounding a VP molecule, determined by hydrogen spin-lattice relaxation technique [61].

It is essential that within the same range of concentration a change of electronic state of the monomer molecule occurs (Chapter 1). Accumulation of solvent (H_2O) molecules in the hydrate shell surrounding a VP molecule affects distribution of electronic charge density on the vinyl group ($CH_2=CH-$) due to conjugation of the latter through a nitrogen atom with a $C=O$ group interacting with water molecules. A significant downfield shift of the $CH_2=$ signal position in the ^{13}C NMR spectrum of VP at limited hydrated surrounding of a monomer molecule by water (10% VP in H_2O) against that of the same signal in pure monomer on 5 ppm demonstrates an appreciable increase of positive charge or decrease of the negative one on this carbon atom (see Table 1.11).

Propagation reaction rate constants (k_p) of VP in water (10%) and in bulk differ greatly being $(2.2 \pm 0.4) \times 10^4$ [100] and $\sim10^3$ l/mol s [101, 102] respectively. The reaction rate constant of biomolecular termination (k_{bt}) of VP monomer in water (10%) is equalled $(6 \pm 2) \times 10^8$ l/mol s [100] and exceeds that in the bulk $(7 \times 10^7$ l/mol s) [102].

It should be taken into account that hydration of a $C=O$ group in a terminal chainlink with the radical shown below, can affect its reactivity relative to the double bond of a monomer molecule. Therefore, the reaction acceleration of *N*-vinylamide polymerization in protic solvents, particularly water, seems to be caused by the increase of reactivity of both the double bond in the monomer molecule and the propagating radical connected with $C=O$ through the nitrogen atom. A NMR technique [61] establishes limiting filling of the hydrated layer shell around a polymer chainlink, and consequently the terminal propagating radical, to be ~15 molecules.

$$\begin{array}{c} \sim\!\!\sim\!\!CH_2 - \overset{\bullet}{C}H \\ | \\ N \\ \diagdown \\ \boxed{}\; C=O\cdots H-O\cdots H-O \\ \qquad\qquad | \qquad\quad | \\ \qquad\qquad H \qquad\quad H \end{array}$$

An appreciable rise of k_{eff} with a decrease of VP concentration ranging from 30 to 10 wt% is caused by two factors: (1) by the increase of the number of water molecules in hydrated shell of the monomer molecule (from 35 to 58),

Table 2.1 The action of the number of H_2O molecules on charge distribution in N-isopropylpyrrolidone

The number of H_2O molecules	Charges on oxygen and carbon atoms	
	$-C=O^a$	$-CH-N=^a$
0	-0.367	-0.0064
2	-0.380	-0.010
5	-0.395	-0.013
10	-0.407	-0.017

^{a} PM-3 method on the AMPAC software.

that brings in a growth of positive charge in the $CH_2=$ group, and (2) by water saturation of the hydrated shell near the propagating polymer radical.

At the accumulation of water molecules close to a $C=O$ group of N-isopropylpyrrolidone simulating the terminal chainlink structure, as is shown by quantum-chemical calculations, there is a growth of negative charge on the oxygen atom in $C=O$ and methine group, connected to the nitrogen atom (Table 2.1).

It is seen (Table 2.1) that at filling of the hydrate shell surrounding the propagating radical ($-CH-\overset{x}{C}H$) a negative charge on the methine carbon atom is enhanced due to the formation of hydrogen bonds of $C=O$ with H_2O molecules. At the same time, hydration of the monomer molecule results in a decrease of negative charge on the $CH_2=$ group of double bond. In turn, it promotes the increase of an effective rate constant of polymerization reaction with a change of VP concentration from 30 to 10 wt% in water. At [VP] < 10%, hydrated layers in close proximity to both the monomer molecule and the terminal polymer chainlink with a radical are saturated by water molecules without change at further dilution. In this case, the order of reaction on the monomer becomes the first order.

N-VINYLACETAMIDE

The unlike behaviour of this monomer in the polymerization reaction with AIBN as a initiator, in comparison with VP, was displayed for protic and aprotic solvents. Akashi and co-workers [39] found different time-conversion curves for the polymerization of N-vinylacetamide in two solvents, indicating that the reaction proceeds two to three times faster in benzene than in ethanol.

This feature of the polymerization reaction may be ascribed to a significant molecular aggregation of monomer molecules in benzene against ethanol due to intermolecular hydrogen bonds between those of *N*-vinylacetamide favouring the propagation step of polymerization. From NMR spectroscopy, when concentration of the monomer increases, the chemical shift of its amide proton in $CDCl_3$ shifts to a lower field owing to hydrogen bond formation.

Here it should be noted that *N*-vinylacetamide molecules in *trans*-conformation associate with dimers and trimers (see quantum-chemical calculations in Chapter 1) and with strong hydrogen bonds and significant dipole moments when these are in the gas phase.

$$CH_2{=}CH \quad CH_2{=}CH \quad CH_2{=}CH$$

The molecular association of monomer molecules in an aprotic benzene solution is likely to accelerate the polymerization reaction, whereas it is destroyed in protic ethanol molecules which are capable of hydrogen bonding monomer molecules.

2.4 MICROSTRUCTURE OF THE POLYMER CHAIN AND CONFORMATIONS OF SIDE RESIDUES

CONFIGURATIONAL SEQUENCES IN POLY-N-VINYL-PYRROLIDONE CHAIN

Microstructure of a PVP polymer chain was studied by ^{13}C NMR methods [103–106]. The synthetic pathways of PVP preparation used did not significantly affect relative contents of various configurational sequences [103]. Thus, PVP samples obtained by radical polymerization in water at 100 °C and in toluene at 85 °C and –78 °C (photo-initiation), provided virtually the same methine carbon signal in the ^{13}C NMR spectra. In the case of samples obtained by cationic polymerization, the relative intensity of carbon signal of the methine group in isotactic configuration slightly increases while that of the group in syndiotactic configuration decreases [104].

It was found [105, 106] that in the ^{13}C NMR spectrum of PVP (aqueous solution) the methine carbon signal split into three lines, which refer to isotactic (mm), syndiotactic (rr), and atactic (mr) sequences (Fig. 2.3).

The $-C_\beta H_2-$ group signal originates from the pentad structure. However, only five lines in the spectrum could be acceptably resolved.

The correlation between configurational sequences in the PVP polymer chain obtained by radical polymerization is detected from the peak ratio of ^{13}C NMR

Table 2.2 Configurational sequences in PVP chain and carbon chemical shift of methine group (aqueous solution) [106]. Reprinted from *J. Polym. Sci. Polym. Lett.* **19**, Cheng, H. M. Smelt, T. E. and Vitus, D. M., 29–31, © 1981, and *Prog. Polym. Sci.* **18**, Kirsh, Yu. E., 519–542, © 1993 with kind permission from Elsevier Science Ltd, The Boulevard, Langford Lane, Kidlington OX5 1GB, UK

| Sequences | Contents (%) | | δ (ppm) |
	From data [103]	From data [106]	in D_2O [106]
mm	24	21	47.8
mr	43	43	46.9
rr	33	36	46.3

signal of $-C_xH-$ group of the chain indicating its atactic microstructure (Table 2.2).

CONFIGURATIONAL SEQUENCES IN POLY-N-VINYLACETAMIDE CHAIN

The 400 MHz ^1H NMR spectra of poly-*N*-vinylacetamide samples obtained by radiation-induced polymerization of the monomer in crystal state at 0 °C and in ethanol solution at 0 °C and also by free radical initiation in ethanol at 60 °C show different intensities of the three methyl proton resonance signals which are assigned to the components of a syndiotactic triad (1.90 ppm), a heterotactic triad (1.93 ppm) and an isotactic triad (1.98 ppm) [39].

Akashi and co-workers [39] showed that poly-*N*-vinylacetamide prepared in solid state has a highly isotactic structure (triad ratio: mm 50%, mr 40%, rr 10%), whereas irradiation-induced polymerization of monomer in ethanol at 0 °C gives an atactic polymer similar to that obtained with participation of free radical initiator in solution (triad ratio: mm 26%, mr 54%, rr 20%). It is possible that molecular aggregation of *N*-vinylacetamide molecules in crystal state through hydrogen bonds creates a suitable structure arrangement for isotactic chain propagation.

CONFIGURATIONAL SEQUENCES IN PVCL CHAINS

The contribution of side substituent structure in the regulation of configurational sequences at free radical polymerization can also be displayed in the case of VCL containing a voluminous seven-membered ring in 'chair' conformation. This ring conformation affects the radical reaction between a terminal chain radical and the double bond of a monomer molecule, since considerable sterical obstacles occur on their mutual approach. The addition of VCL to a propagating chainlink is most suitable in syndiotactic sequences in a polymer chain where the obstacles are minimal. In fact, the carbon atom signal of the methine group

(a)

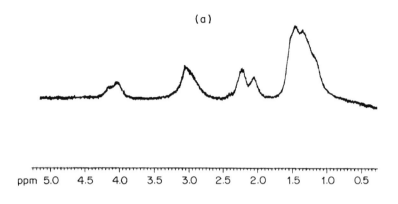

ppm 5.0   4.5   4.0   3.5   3.0   2.5   2.0   1.5   1.0   0.5

(b)

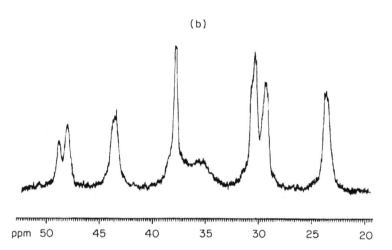

ppm 50   45   40   35   30   25   20

Fig. 2.4. The ^{1}H (a) and ^{13}C (b) NMR spectra of poly-*N*-vinylcaprolactam in a D$_2$O solution. (Bruker AMX-400).

in the ^{13}C NMR spectrum of PVCL obtained by free radical polymerization with AIBN in isobutanol consists of two peaks not of three as is characteristic of atactic microstructure (Fig 2.4).

Proton resonance of the same group in ^{1}H NMR spectrum of PVCL in D$_2$O also gives two peaks (Fig. 2.4) The position of peaks in these spectra is sure proof of the presence of triads of two types in a polymer chain [44, 49] indicating, in principle, the syndiotactic configuration.

The appearance of syndiotactic microstructure in a PVCL main chain prepared by radical polymerization of VCL becomes apparent if the structural arrangement of monomer molecules in a unit cell of the crystal is considered. In the unit cell two VCL molecules (in 'chair' conformation) are brought close together so that vinyl groups are arranged far apart from each other as oppositely directed 'backs' of two 'chairs' which are placed on one another by 'seats'. One can assume that the formation of associates of the structure similar to that of VCL in the unit cell after the dissolving of VCL crystals is due to large dispersion and dipole–dipole interactions between rings. This association could happen between a terminal propagating radical and these associates. Since they are voluminous and involve two vinyl groups which are separated as widely as possible, the approach of the vinyl group of these associates of a propagating terminal radical is restrained by sterical abstacles. A minimum force of repulsion between voluminous associates and a terminal chainlink seems to be achieved through syndiotactic sequence formation in a VCL polymer chain.

Hence, these two examples point out the fact that the structure of side substituents in chain propagation reactions at the radical polymerization of N-vinylamides affects the microstructure of poly-N-vinylamide chains. In the case of VCL molecules being in 'chair' conformation, considerable sterical obstacles originating from the approach of a monomer molecule to a terminal chain radical regulate their certain arrangement in the propagation reaction, bringing about syndiotactic configurational sequences. Proton bonding interactions in crystals of N-vinylacetamide seem to be responsible for isotactic sequence formation of polymer chains at radiation-induced polymerization.

CONFORMATIONAL TRANSFORMATIONS OF ACYCLIC SIDE SUBSTITUENTS IN CHAIN PROPAGATION REACTION

N-vinylamides of aliphatic carboxylic acids in a solution consist of two conformers. Their contents depend on the structure of the substituent at the nitrogen atom (see Chapter 1). For N-vinylamides with a hydrogen atom as a substituent the contents of *trans*-form is predominant, whereas in N-vinylamides with alkyl groups the contents of *cis*-form appreciably exceeds that of *trans*-form (Table 1.6).

The conformational state of these side amide groups at radical polymerization reaction (solvent, 60 °C, AIBN) is much transformed as is shown in refs [47, 48]. The ratio of conformers in a chain differs from that in monomers, which is confirmed by, e.g. ^{13}C NMR spectra of poly-N-vinyl-N-methylformamide and poly-N-vinyl-N-methylacetamide (Fig. 2.5).

The existence of conformers of different contents in a side substituent in poly N-vinylamides is proved by a distinct resolution of proton signals in the —C$_x$H— group of a chain or in the CH$_3$—N< group on to two peaks in the ^{1}H NMR spectra or that of carbon atom in the —C$_x$H— group in the ^{13}C NMR spectra (Table 2.3).

In the case of poly-N-vinyl-N-methylformamide the intensity of the two peaks of the —C$_x$H— and CH$_3$—N< groups is identical (50 : 50) (Fig. 2.5),

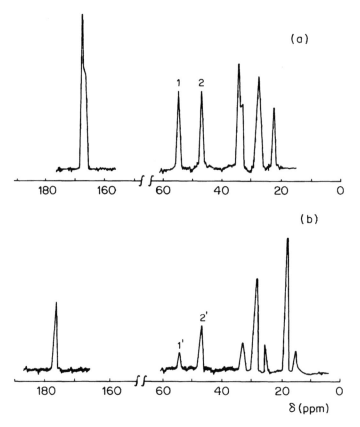

Fig. 2.5. The ^{13}C NMR spectra of poly-*N*-vinyl-*N*-methylformamide (a), and poly-*N*-vinyl-*N*-methylacetamide (b) in water [47, 106]. Reproduced by permission of Nauka.

indicating the identical ratio of the two conformers in the side substituent. The structure of poly-*N*-vinyl-*N*-methylformamide could be presented as a chain of 'alternating' copolymer.

It is interesting that the *trans*-form prevails (~ 90%) in the structure of the side substituent in all other poly-*N*-vinylamides investigated (Tables 2.3 and 2.4).

Such an assignment of the maximum intensity peak in the ^{13}C NMR spectrum of the —C$_\alpha$H— group, being upfield, is performed by comparative analysis of

Table 2.3 The ^{1}H and ^{13}C NMR chemical shifts of poly-N-vinylamides in D_2O [47, 48]. Reproduced by permissin of Nauka

Polymer	—C$_x$H—		CH$_2$—(chain)		—N—CH$_3$		—N—CH$_2$—	
	^{1}H	^{13}C	^{1}H	^{13}C	^{1}H	^{13}C	^{1}H	^{13}C
PVMF	3.89	55.3 (50%)	2.02	35.0	3.49	26.9 (50%)		
	4.65	47.1 (50%)			3.40	30.9 (50%)		
PVA	3.9		1.6					
PVMA	3.85	55.1 (10%)	1.94	36.4				
	4.46	48.7 (90%)						
		54.7 (5%)						
PVEA	4.59	48.0 (95%)	2.12	37.6			3.34	38.5
		54.5 (10%)						
PVEP	4.70	47.7 (90%)	1.95	38.2			3.52	38.2
PVPA	4.60	54.3	1.96	36.9			3.50	36.2
		47.5						

Notes: The values of δ (^{1}H) of the CH$_3$ group in —N—CH$_2$—CH$_2$—CH$_3$— and $-\overset{\overset{\text{O}}{\|}}{\text{C}}$—CH$_2$—CH$_3$ are equal to 1.34 and 1.34 ppm respectively; the value of δ (^{13}C) in the same groups –12 and 10.4 ppm, δ (^{1}H) in group $-\text{C}\overset{\nearrow \text{O}}{\underset{\searrow \text{H}}{}}-$ 8.59 ppm and in group $-\overset{\overset{\text{O}}{\|}}{\text{C}}-\overset{\bullet}{\text{C}}\text{H}_2-$ is 2.63 ppm; the values of δ (^{13}C) in $-\overset{\overset{\text{O}}{\|}}{\text{C}}-\overset{\bullet}{\text{C}}\text{H}_2-$ is 16.8 ppm.
Abbreviations: PVMF poly-N-vinyl-N-methylformamide, PVA poly-N-vinylacetamide, PVMA poly-N-vinyl-N-methylacetamide, PVEA poly-N-vinyl-N-ethylacetamide, PVEP poly-N-vinyl-N-ethylpropionamide, PVPA poly-N-vinyl-N-propylacetamide.

Table 2.4 Conformation contents in side residues of poly-N-vinylamides (by ^{13}C NMR data) [47, 48]. Reproduced by permission of Nauka

Polyamide	Trans-form content (%) at various temperatures		
	25 °C	60 °C	95 °C
Poly-N-vinylformamide	~100	~90	85
Poly-N-vinylacetamide	100	~100	100
Poly-N-vinyl-N-methylformamide	50	50	50
Poly-N-vinyl-N-methylacetamide	93	90	80
Poly-N-vinyl-N-ethylpropionamide	93	—	83
Poly-N-vinyl-N-propylacetamide	90	85	—

Note: Conditions of measurement: 30% water solutions, error of measurement ± 3%.

the position of this peak and that in the case of PVP, where this group is in *cis*-position to C=O, i.e. VP is 100% in *trans*-form due to the cyclic structure of the pyrrolidone ring. So, the value of δ(^{13}C) for the signal of —C$_x$H— group in PVP is in the range of 46.6 to 47.8 ppm that almost coincides with the chemical shift of high intensity signals ($\delta \approx$ 47.1–48.0 ppm) for poly-N-vinylamides of aliphatic carboxylic acids (Table 2.4). It proves dominating contents of the *trans*-form in these poly-N-vinylamides. As follows from the reference data [107], in each pair of signals the upfield peak is attributed to a carbon atom, being in *cis*-position to a C=O group. As a result, the peak of the maximum intensity signal of the —C_αH— group corresponds to the *trans*-conformer, whereas the low-intensity one corresponds to the *cis*-conformer.

The contents of the *trans*-form of the side substituent in these poly-*N*-vinyl-amides are represented in Table 2.4.

The increase of temperature of polymer solutions slightly reduces *trans*-form contents from 90–100% to 80–85%. Only in the case of poly-*N*-vinyl-*N*-methylformamide does the ratio of the two conformations remain constant.

The comparison of conformational states of monomers and polymers obtained from these monomers (Tables 1.6 and 2.4) shows the great difference between them.

N-vinylformamide monomer being 70% in *trans*-form is polymerized with the formation of polymer being ~100% in *trans*-form. The conformational state of the other monomers (~30% in *trans*-form), except for *N*-vinyl-*N*-methyl-formamide, changes at the chain propagation reaction resulting in formation of a polymer where the side substituent is 90–95% in *trans*-form. However in the case of *N*-vinylacetamide and poly-*N*-vinylacetamide the contents of the *trans*-form do not change, remaining at 100%. And finally, poly-*N*-vinyl-*N*-methylformamide, unlike these two groups of polymers, contains identical quantities of *trans*- and *cis*-forms (50:50).

These results testify that an essential reorganization of the side substituent of a terminal propagating chainlink occurs during reaction of chain propagation. In our opinion, the major factor regulating the conformational state at the polymer-ization process, is likely to be the interaction between side substituents of the terminal chainlink on a chain and those of the monomer molecule. As follows from quantum-chemical calculations (Table 1.8), the energy of monomer associ-ation in a solution differs markedly, depending on the monomer conformational state. So, for *N*-vinylacetamide the heat of associatiion (ΔH) of three molecules in a *trans-trans-trans*-conformer aggregate is the greatest (-35.7 kJ/mol). Therefore the *trans*-form prevails in a poly-*N*-vinylacetamide chain.

At *N*-vinyl-*N*-methylformamide polymerization the side residue of a terminal chainlink co-operates with the same group of a monomer molecule due to dipole–dipole forces, which are most strong for a pair of *cis-trans*-conformers (Table 1.8). Therefore, the pairs in *cis*- and *trans*-forms form a polymer of 50:50 conformer contents.

One should note the effect of the CH_3— group incorporation instead of the hydrogen atom into carbonyl of monomer on the conformational transition in side substituents at radical polymerization in going from *N*-vinyl-*N*-methyl-formamide to *N*-vinyl-*N*-methylacetamide. In fact, there is an alternative con-formational set in the side residues of poly-*N*-vinyl-*N*-methylacetamide in comparison to poly-*N*-vinyl-*N*-methylformamide. Quantum-chemical calculations (Table 1.8) indicate that the association of two monomer molecules such as N-vinyl-N-methylacetamide is most favourable energetically in the case of *trans-trans* conformer association, in a similar way that it occurs at the monomer molecule approaching a terminal radical at a chain propagation reaction.

Thus, the monomers of the first group with hydrogen atom at nitrogen atom form polymers where side residues are in *trans*-form, as these monomers contain the greater portion of the latter. In addition, the terminal chainlink can interact with the monomer molecule, forming a hydrogen bond between the   NH and C=O groups.

Monomers of the second group with alkyl substituents at the nitrogen atom and at the C=O group also enter the chain in *trans*-form because of dipole–dipole interactions. Voluminous methyl groups seem to create sterical obstacles to the approach of the monomer molecule in *cis*-form to a terminal radical chainlink at the formation of a polymer chain.

Only the monomer with the CH_3— group at nitrogen atom and hydrogen atom at the C=O group, namely N-vinyl-N-methylformamide, construct the chain with alternation of *cis*- and *trans*-forms due to favourable dipole–dipole interactions.

Summarizing this section it is necessary to note that the conformational structure of side residues in poly-*N*-vinylamides is unlike that of appropriate monomers and is defined by the energy of association of a terminal chainlink and a monomer molecule with the participation of their side residues controlling the contents of conformers in *cis*- or *trans*-forms in polymer.

2.5 POLYMERIZATION IN THE PRESENCE OF HYDROGEN PEROXIDE

The research of the mechanism of MW regulation of medicinal poly-*N*-vinylpyrrolidone, traditionally [2, 11, 67, 69] prepared with the help of hydrogen peroxide, was performed by studying VP polymerization reaction with the application of well-known methods of initiation (UV-initiation or thermal initiation with AIBN) in organic or aqueous solutions [81, 100, 108].

UV-INITATION: HYDROGEN PEROXIDE IN WATER

In an aqueous solution involving a monomer and H_2O_2, under the action of UV-light of $\lambda = 313$ nm the radicals are formed due to H_2O_2 degradation [100]. These radicals can interact with double bonds both in the monomer and H_2O_2. The reaction rate constants are equal to 7×10^9 1/mol s and $\sim 3 \times 10^7$ 1/mol s respectively. Therefore, when VP concentration considerably exceeds that of hydrogen peroxide, $HO^{\bullet}$, formed in the system, reacts only with monomer. The subsequent reaction of $^{\bullet}OH$ (2.11) and VP gives a primary radical, initiating the further polymerization (2.11).

(2.11)

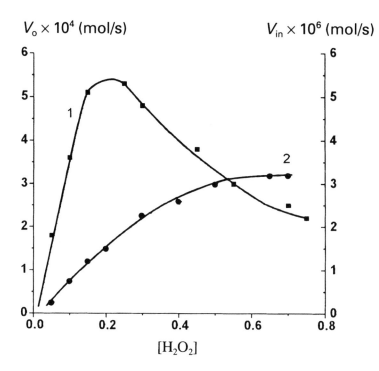

Fig. 2.6. Dependence of initial polymerization rate for VP (V_o) (1), and rate of initiation reaction (V_{in}) (2) on hydrogen peroxide concentration, UV-light with $\lambda = 313$ nm, [VP] = 1.0 mol/l, without O_2, H_2O [78]. Reproduced by permission of Nauka.

Shtamm and co-workers established [100] that at small [H_2O_2] providing the linear dependence of the initiation reaction rate (V_{in}) on light intensity (J_o), the dependence $V_{in} \sim J_o^{0.5}$ is observed, indicating the prevailing square-law chain termination reaction. The absence of agreement on the behaviour of V_o and V_{in} from [H_2O_2] (Fig. 2.6) is a confirmation of H_2O_2 participation in linear termination of polymer chain at [H_2O_2] > 0.2 mol/l. As seen in Fig. 2.6, the increase of [H_2O_2] causes a rise of the initiation reaction rate, reaching a constant value at [H_2O_2] > 0.8 mol/l, whereas the initial polymerization reaction rate at first grows with the increase of [H_2O_2], then achieves a maximum at [H_2O_2] = 0.2 mol/l and decreases. Otherwise the polymerization reaction rate would have changed proportionally to V_{in}. The propagating radical attacks hydrogen peroxide molecule and abstracts the hydrogen atom (2.11).

$$\sim R^\bullet + H{-}O{-}O{-}H \xrightarrow{k_{l.ter}} \sim R{-}H + HO_2^\bullet \qquad (2.12)$$

In the absence of O_2 when the linear termination of a chain (at [H_2O_2] > 0.2 mol/l) becomes dominating, the kinetic equation 2.2 of the polymerization reaction is as follows:

$$V = k_p \,[\text{VP}][\text{R}] = k_p V_{in}[\text{VP}]/k_{1.ter}\,[\text{H}_2\text{O}_2] \qquad (\text{eq. 2.2})$$

where k_p and $k_{1.ter}$ are reaction constants of chain propagation and linear termination. The equation 2.2 determines $k_{1.ter}$ and the ratio of $k_p/k_{1.ter}$ to be 10^2 mol/l s and 67.5 ± 0.5 respectively [81, 100].

THERMAL INITIATION: HYDROGEN PEROXIDE IN ORGANIC SOLVENT

Karaputadze and co-workers [81] found that the dependences of the initial reaction rate (V_o) of VP polymerization in isopropanol on initiator concentration (AIBN) in the presence and in the absence of H_2O_2 differ very much (Fig. 2.7).

The reaction rate, as seen in Fig 2.7, greatly decreases with the introduction of H_2O_2. The order of reaction on the initiator in the absence of H_2O_2 is 0.5, whereas in the presence of 0.07 mol/l H_2O_2 it is increased up to 1. The different order of reaction on initiator concentration reflects a change of the chain termination reaction character with the increase of $[H_2O_2]$ from the square law in the absence of H_2O_2 to the linear at high $[H_2O_2]$.

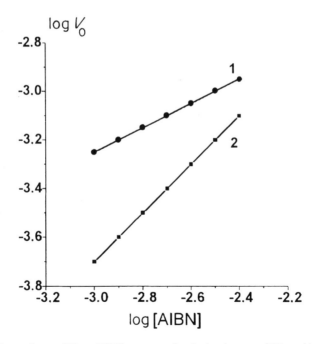

Fig. 2.7. Dependence of V_o on AIBN concentration in the absence of (1), and in the presence of H_2O_2 (2) [81]. Reaction conditions: 75 °C, isopropanol, [VP] = 1 mol/l, $[H_2O_2]$ = 0.07 mol/l [81]. Reproduced by permission of Nauka.

Analytically the effect of H_2O_2 on the VP polymerization reaction rate is described by the equation 2.3.

$$V_{in} = k_{1.ter}[R^\cdot][H_2O_2] + 2k_{sq.ter}[R^\cdot]^2 \qquad \text{(eq. 2.3)}$$

where $V_{in} = 2k_{in}[AIBN]$ is the rate of initiation reaction, $[R^\cdot] = V_0/k_p$ [VP] the stationary concentration of chain propagating radicals; k_{in}, $k_{1.ter}$, $k_{sq.ter}$ and k_p the rate constants of the reactions of initiation, linear and square-law termination and propagation of chain respectively.

In the case of linear termination reaction predominance the proportional dependence of V_0 on V_{in} is observed and the contribution of the second item in the last equation is negligibly small. Then the new equation 2.4 is written out in the following form:

$$V_0 = k_p V_{in}[VP]/k_{1.ter}[H_2O_2] \qquad \text{(eq. 2.4)}$$

As shown in ref. [79], the dependence of V_0 on $[H_2O_2]^{-1}$ is linear, confirming the linear chain termination reaction with the participation of H_2O_2. As a result, the increase of H_2O_2 concentration in VP solutions brings about the reduction of PVP molecular weight (M_n) (Fig. 2.8). Thus, in an organic solvent (isopropanol) H_2O_2 is a regulator of polymer MW, participating in the linear termination of a chain.

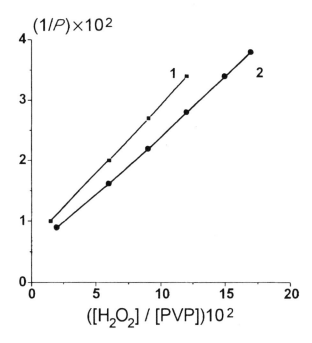

Fig. 2.8. Dependence of polymerization degree (P) for PVP at constant V_0 in isopropanol (1) and water (2) on mole ratio of $[H_2O_2]/[VP]$. Conditions: 75 °C, 10 wt%, AIBN [81]. Reproduced by permission of Nauka.

THERMAL INITIATION: HYDROGEN PEROXIDE IN WATER

As shown in ref. [81], the order of reaction on the initiator (AIBN) in aqueous solution as well as in isopropyl alcohol in the absence of H_2O_2 is equal to 0.5 due to the bimolecular character of the termination reaction. However, the introduction of H_2O_2 to an aqueous reaction mixture involving VP fails to increase the order although the growth of the latter is observed for organic solvent. The considerable drop of the reaction order on AIBN is displayed. In fact, the dependence of V_o on $[H_2O_2]$ in water also differs from the observed in the case of isopropanol; the effect of reaction inhibition is expressed in a much smaller degree [81].

It is well known [79], that the mechanism of reduction–oxidation reaction with the participation of H_2O_2 in aqueous solutions depends on trace impurities of transition metals, predominantly iron ions.

It is found [81], applying carefully purified water and water containing a controlled quantity of iron ions, that the order of reaction on H_2O_2 depends on $[Fe^{3+}]$. So, at $[Fe^{3+}] = 10^{-8}, 10^{-7}, 10^{-6}$ and 3×10^{-5} mol/l, the order of reaction on $[H_2O_2]$ is $-0.5, -0.3, -0.2$ and 0 respectively. At $[Fe^{3+}] > 3 \times 10^{-5}$ mol/l the reaction rate becomes independent from $[H_2O_2]$. In an ideal situation of complete absence of iron ions in the reaction medium (after careful purification of all initial reagents) the action of H_2O_2 should be the same, as of isopropyl alcohol. However, such a situation is impossible in aqueous solutions.

The participation of H_2O_2 in chain termination reaction (2.12) results in the formation of HO_2^* radicals in a solution. The interaction of HO_2^* or O_2 with trace Fe^{3+} impurities generates Fe^{2+} ions in the system and promotes reinitiation of the VP polymerization process. In the presence of iron ions the interdependent processes of VP polymerization and catalytic H_2O_2 decomposition with the formation of hydroxyl radicals are realized (reactions 2.13–2.18).

$$R \sim CH_2 - \overset{\bullet}{C}H \qquad\qquad R \sim CH_2 - CH_2$$

$$\underset{\text{+ } H_2O_2}{\overbrace{}N\overset{O}{\underset{C}{\diagup\!\diagdown}}} \xrightarrow{\ k_{1.ter}\ } \underset{\text{+ } HO_2^{\bullet}}{\overbrace{}N\overset{O}{\underset{C}{\diagup\!\diagdown}}} \qquad (2.15)$$

$$HO_2^{\bullet} + -\overset{|}{\underset{|}{Fe}}{}^{3+} \longrightarrow -\overset{|}{\underset{|}{Fe}}{}^{2+} + O_2 \uparrow + H^+ \qquad (2.16)$$

$$-\overset{|}{\underset{|}{Fe}}{}^{2+} + H_2O_2 \longrightarrow -\overset{|}{\underset{|}{Fe}}{}^{3+} + HO^{\bullet} + HO^- \qquad (2.17)$$

$$CH_2 = CH \qquad\qquad HO - CH_2 - CH - CH_2 - \overset{\bullet}{C}H$$

$$HO^{\bullet} + \overbrace{}N\overset{O}{\underset{C}{\diagup\!\diagdown}} \xrightarrow{\ k_p\,(VP)\ } \overbrace{}N\overset{O}{\underset{C}{\diagup\!\diagdown}} \quad \overbrace{}N\overset{O}{\underset{C}{\diagup\!\diagdown}} \qquad (2.18)$$

So, hydrogen peroxide at VP polymerization reaction in organic and aqueous solutions participates in a degenerate transfer reaction of a chain, influencing the polymerization reaction rate and the MW of obtained polymers. It was also found that for the other N-vinylamides (VCL and N-vinyl-N-methylacetamide) the introduction of H_2O_2 in reaction mixtures causes a decrease in MW [75,108].

As polymerization reactions with the participation of hydrogen peroxide have been presented above, it is now essential to focus upon the mechanism of VP water polymerization in the traditional system involving peroxide and ammonia [67, 69] and the factors regulating the polymerization rate, MW and MW distribution of the obtained polymers and causing secondary reactions, modifying polymer chain and deteriorating product quality [76, 82, 108–110].

Hydrogen peroxide participates in a number of reactions. They are:

1. Reduction–oxidation reactions of initiation with trace impurities of iron ions;
2. Linear chain termination reaction;
3. Oxidation of ammonia and other organic impurities present in a monomer;
4. Hydrolysis reaction of lactam rings.

H_2O_2 concentration in a solution relative to that of a monomer dictates the quality of polymer obtained, namely its MW and MW distribution. H_2O_2 participation in various reactions changes the concentration ratio ($[[H_2O_2]/[VP]$]) resulting in widening of MW distribution up to 3.5–4.0 [109]. Fast decomposition of H_2O_2 reduces a polymer yield, requiring residual monomer extraction by organic solvent from a polymer solution [2].

It should be noted that ammonia oxidation reaction can originate in the presence of H_2O_2. Hydrazine, as it is known [69], is synthesized from ammonia with the participation of oxidizing agents. It is possible that in the presence of H_2O_2, iron ions and ammonia in the system HO˙ radicals are formed reacting with ammonia (2.18, 2.19) and creating toxic hydrazine

$$NH_3 + \text{˙OH} \longrightarrow \text{˙NH}_2 + H_2O \tag{2.19}$$

$$H_2N\text{˙} + H_2N\text{˙} \longrightarrow NH_2-NH_2 \tag{2.20}$$

$$2\text{˙NH}_2 \longrightarrow N_2H_4 \tag{2.21}$$

As industrial PVP can contain traces of hydrazine [111], there is a limitation on the amount of hydrazine in medicinal PVP powder (no more than 1 part per million) in US pharmacopoeia (USP 1985, FAO 1986).

In a number of reports [2, 11, 70, 71, 108, 109] it is noted that low molecular weight polymers of *N*-vinylpyrrolidone prepared in the traditional manner lower the value of pH of pure water to 3.7–5.0 after dissolving. It indicates the presence of substances with acidic groups in polymer products, the dissociation of which reduces pH. The occurrence of carboxylic groups was assumed previously to be connected to the formation of acetic acid due to acetaldehyde oxidation [2, 112, 113].

It was found, however, [82], that PVP fractions obtained in the presence of H_2O_2 contain chain fragments with a carboxylic group. The fractionation of a PVP sample ($M_w = 9.6 \times 10^3$) on G-200 Sephadex and the additional polymer fractions processing in water by passing through anion-exchange and cation-exchange resins do not free all fractions from an acidic component. PVP fraction macromolecules of M_w ranging from 5×10^3 to 2.5×10^4 ($M_w/M_n = 1.8$) contain some two to three carboxylic groups per 10^3 chainlinks regardless of fraction MW. Potentiometric titration displayed the value of pK_a for an acidic group to be ~ 5.0 [82]. In this case the polymer chain apparently contains chainlinks with a carboxylic group of the following structure:

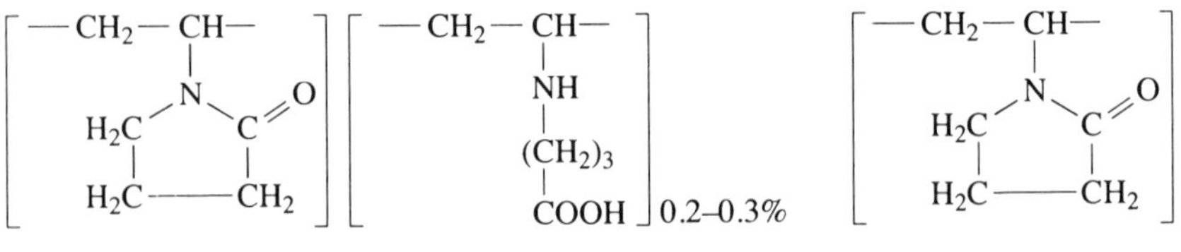

The production of hydrolysed side residues during VP polymerization in the presence of H_2O_2 in alkaline medium (pH = 8–10) could be caused by the fact that nucleophilic groups of HOO^- type (pK_a of H_2O_2 = 4.75) present in the solution because of H_2O_2 dissociation catalyse the hydrolysis reaction of the amide group in PVP lactam ring. It is known [114] that HOO^- is the most potent nucleophilic catalyst of hydrolysis reaction of esters and amides.

Thus, VP polymerization in the presence of H_2O_2 and NH_3, though apparently simple to be realized, is a rather complicated process requiring precise control over the quality of all reagents and performance conditions. Some side reactions can occasionally appear for unknown reasons, deteriorating the quality of polymer products. Therefore, over the last 10–15 years a large number of works was dedicated to the search of new pathways of low molecular-weight PVP preparation without the drawbacks mentioned above [71–76, 82, 83, 96, 97, 108, 115].

2.6 NON-TRADITIONAL POLYMERIZATION PROCESSES

ALIPHATIC HYDROPEROXIDES AND PEROXIDES

A series of compounds is used instead of hydrogen peroxide in order to control the MW of poly-N-vinylamides at VP polymerization in organic solvents [71, 72, 81, 83, 115]. These are *tert*-butyl hydrogen peroxide, cumyl hydrogen peroxide, di-*tert*-butyl peroxide and other compounds participating in the polymerization process not only as MW regulators, but also as initiators. Similar to hydrogen peroxide, a molecule of organic hydrogen peroxide can enter into the reaction of linear chain termination reducing polymer MW (reaction 2.22).

$$\text{(2.22)}$$

The introduction of *tert*-butyl hydrogen peroxide as initiator in a mixture of VP (15 wt%) + isopropanol (85 wt%) was found to decrease [η] of PVP samples from 0.14 to 0.07 dl/g on the increase of initiator concentration from 1 to 8 wt% in relation to monomer [71, 115].

The majority of the patents [71–75, 83] recommend isopropyl alcohol as an organic solvent for VP polymerization process as it does not lower reaction rate and reduces polymer MW acting as a chain transfer agent.

The ^{1}H and ^{13}C NMR techniques [116] indicate the presence of terminal fragments both of isopropanol (solvent) and of AIBN (initiator) when VP polymerization is carried out in dilute isopropanol solutions of VP in a concentration range from 2.0 to 15.0 wt% of the monomer and AIBN concentration ranging from 0.5 to 3.0 wt%.

$$\text{(2.23)}$$

$$\text{(2.24)}$$

$$\text{(2.25)}$$

$$\text{(2.26)}$$

Varying the ratio [VP]/[isopropanol] it is possible to synthesize PVP of MW from 3×10^3 to 6×10^4 with hydroxyl and nitryl terminal groups.

Aliphatic peroxides (di-*tert*-butylperoxide, di-cumylperoxide and others) are used as initiators of VP polymerization at high temperature (from 120 to 200 °C) [6, 115]. In order to decrease their decomposition temperature one should apply salts of heavy metals such as copper acetate, copper stearate and others. In this case polymerization reaction is conducted at 80–100 °C. The application of aliphatic peroxides does not allow one to obtain a PVP of low MW.

As reported in [111], the low-molecular-weight medicinal PVP is fabricated in industry not only by VP polymerization in water with H_2O_2 and NH_3 but also in isopropanol, probably with aliphatic peroxides.

RADICAL POLYMERIZATION OF WATER-INSOLUBLE N-VINYLCAPROLACTAM IN WATER

Eisele and Burchard [86] recently studied the radical polymerizaton reaction of VCL in water at 50–70 °C using a well-known initiator (AIBN). The performance of VCL radical polymerization in these conditions is complicated by poor solubility of the monomer in heated aqueous solution and the immediate precipitation in the separated phase of the polymer obtained

In order to avoid these difficulties, VCL polymerization reaction is carried out in water in the presence of a surface-active substance (SAS) as an emulsifier, namely of Aerosol OT (sodium 1,2-bis(2-ethylhexyloxycarbonyl)-1-ethanesulfonate). A large amount of emulsifying agent (from 20 to 100 wt% relative to monomer concentration) is used at creating stable and transparent microemulsions of VCL in water at 50–70 °C. Variation of the concentration ratio of monomer and initiator (or emulsifying agent) allows one to prepare PVCL samples of various MW. The values of M_w are fairly high, ranging from 7×10^5 to 1.2×10^6. PVCL samples obtained by microemulsion polymerization were found to have a wide molecular weight distribution ($M_w/M_n = 3.5$–4.3).

It should be noted that at preparation of pure samples of these polymers an additional treatment of polymer solutions is required with the use of anion-exchange resin (Amberlyt A-26) removing ions of SAS from the polymer solution [86].

VP RADICAL POLYMERIZATION AND RING-OPENING POLYCONDENSATION

The method of preparing VP polymer products insoluble in water, acids and organic solvents is described in ref. [117]. It is based on heating of the monomer under vacuum (100 mm Hg) in the presence of catalysts such as alkaline metals, their oxides or hydroxides in the amount of 0.1–0.5 wt% with respect to monomer concentration. The reaction is performed in a temperature range from 150 to 180 °C. The same catalysts were used at preparing insoluble products of VP by heating VP + water mixture at 140 °C in autoclave [118].

Hofman and Herrle [119] pointed out that the performance of the polymerization process by the method mentioned above [118] results in the preparation of coloured polymer products. They developed a new pathway of synthesis of colourless PVP compounds insoluble in water and in other solvents. VP polymerization was carried out in the presence of crosslinking reagents such as N,N-divinylimidazolidon-2, N,N-divinylhexahydropyrimidine-on-2 and other and well-known initiators (AIBN, organic peroxides and so on). A crosslinked PVP product fabricated by this method was proposed for use in the purification of beer, wine and juices.

It should be noted that the information on the structure of VP polymerizates produced in the presence of alkaline metals is absent in refs [117, 118] as well as on the mechanism of non-traditional polymerization reaction. At the same time insoluble crosslinked PVP products find widespread application in the pharmaceutical and food industries. These products named polyvinylpolypyrrolidone (PVPP) are fabricated by BASF (Kollidon CL and Devirgan trade marks) and ISP (USA) (Polyplasdone XL and Polyclar trade marks).

The elucidation of the mechanism of the PVPP production process in the absence of radical polymerization initiators is hampered as the appropriate information is unavailable, being the 'know-how' of the firms concerned. At the same time the establishment of chemical reactions e.g. so-called popcorn polymerization, resulting in the formation of crosslinked polymers without crosslinking agents is of great interest in poly-N-vinylamide chemistry.

A rough idea of PVPP structure is given by IR spectroscopic measurement of commercial POLYCLAR-10 (ISP product for food industry). At present extensive information, concerning IR-spectral characteristics of PVP [120] and other poly-*N*-vinylamides containing aliphatic substituents [20, 121, 122] is available, allowing conclusions to be drawn about the PVPP structure.

It is of interest to consider the IR-spectra of PVP, PVCL, poly-*N*-vinyl-*N*-methyl acetamide (PVMA), poly-*N*-vinyl-*N*-propylpropionamide (PVPA) and crosslinked PVPP (POLYCLAR-10). The former four polymers were prepared in the form of films and the latter (PVPP) was made as finely divided powder in oil. Only one band with the absorption maximum in a range of frequencies ($v_{C=O}$) from 1700 to 1620 cm^{-1} is observed in IR spectra of polymers (films), while two broad bands appear at $v_{C=O} = 1690$ and 1620–1630 cm^{-1} in the case of PVPP. The ratio of band intensities at these frequencies is approximately 60 : 40.

The band of $v_{C=O} = 1690$ cm^{-1} corresponds to the strenghening vibrations of C=O in a strained pyrrolidone ring, whereas the bands at $v_{C=O} = 1630$ (PVCL), 1650 (PVMA) and 1625–1630 cm^{-1} (PVPA) are due to the strengthening vibration of C=O in an unstrained ring of PVCL [120, 122] and in acyclic amides [20, 60]. Comparison of the IR spectra findings enables one to relate the band at $v_{C=O} = 1620$–1630 cm^{-1} in the PVPP IR spectrum to the strengthening vibrations of C=O, being in acyclic open-cut amide groups which originate at the destruction of the pyrrolidone ring at heat processing of VP.

In the crosslinked PVPP about 60% of pyrrolidone rings are probably preserved intact and 40% are open-cut. The rigid construction of a very strong crosslinked PVP system is confirmed by a very slight degree of swelling in water (about 10%).

It seems that VP polymerization reaction under severe conditions proceeds simultaneously through the two pathways (at close reaction rates) namely through radical heat polymerization of vinyl groups and through polycondensation of cyclic amides in the presence of a catalytic amount of water. It is well known [16] that lactams (pyrrolidone or caprolactam) are polymerized in polyamides at high temperature (150–220 °C) with water as a catalyst. High temperature can result in radical polymerization of vinyl containing monomer.

One can assume also that water molecules are changed under the action of polar C=O in amides with enhancement of the basicity of its oxygen atom and of the acidity of its protons (see Chapter 3). In this case the changed water molecule probably possesses increased catalytic activity in the ring-opening reaction, lowering the polycondensation process temperature. Mole concentration ratio of VP and water in reaction medium could be of fundamental importance in controlling the rates of the two reactions (radical polymerization and ring-opening polycondensation).

Thus the structure of PVPP can be represented as follows:

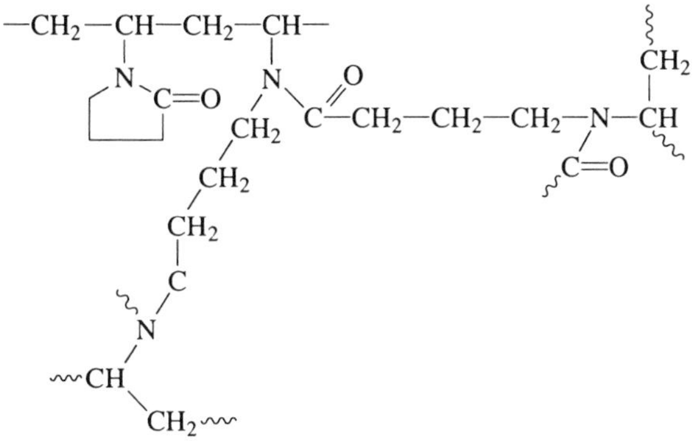

Radiation-induced Polymerization

The radiation-induced (γ-radiation of ^{60}Co) polymerization of VP in water and in various alcohols is studied by a calorimetric method [94]. The effect of VP concentration in these mixtures on the radiation-induced polymerization reaction rate was established to be similar to that of the monomer at different initiation reactions (UV-light or AIBN) at the preparation of PVP samples of high MW ranging from 4×10^5 to 1.5×10^6. It is essential that the samples obtained at various absorbed doses in alcohol (ethanol or butanol) have a fairly narrow MW distribution at complete conversion of monomer to polymer, namely $M_w/M_n = 1.7 \pm 2.7$. It is the author's opinion that this pathway is most promising at development of a high-quality PVP technology.

2.7 COPOLYMERIZATION

N-VINYLPYRROLIDONE AND N,N-DIMETHYL-N,N-DIALLYL-AMMONIUM CHLORIDE (DMAA)

The interesting peculiarity of copolymerization reaction of these two monomers is displayed in the comparison of copolymer composition and monomer contents in a solution [123, 124]. The composition of copolymers, determined with the help of several independent methods at all mole ratios, is strictly consistent with that of monomers in water or alcohol. In this case monomer reactivities at copolymerization reaction are equal to one.

$$r_1^{\text{eff}} = r_2^{\text{eff}} = 1 \quad \text{or} \quad k_{11} = k_{22} = k_{12} = k_{21}$$

The azeotropic copolymerization ability of the pair of monomers is kept at high degrees of converson in both water and ethanol at monomer concentration in a range from 1–5 mol/l and 35–80 °C [123]. In other words, it is possible to fabricate copolymers of this type with a high yield. The composition of these copolymers is identical to that of the monomers in solution.

$$\left[\begin{array}{c} -CH_2-CH \\ | \\ N \\ \diagdown C{=}O \end{array}\right]_x \left[\begin{array}{c} -CH_2-HC-\!\!-\!\!CH-CH_2- \\ \diagdown\overset{+}{N}\diagdown C \quad Cl^- \\ CH_3 \diagup \diagdown CH_3 \end{array}\right]_y$$

CP VP–DMAA

The established phenomenon seems to be caused by structural conformity of terminal propagating chainlinks having a five-membered ring in plane conformation.

The role of the structural factor is observed at copolymerization reaction when each of the monomer pairs considered above is replaced respectively by another monomer of close structure, but containing more voluminous substituents. So, *N*-vinylcaprolactam is applied instead of VP and *N,N*-dimethyl-*N,N*-dimethyl-*N,N*-diallylammonium chloride is replaced by *N,N*,-diallyl-*N*-methyl-*N*-carbisopropyloxymethylammonium chloride (DAMP) [123].

$$\begin{array}{cc} CH_2{=}CH \quad CH{=}CH_2 \\ | \qquad | \\ H_2C\diagdown \quad \diagup CH_2 \quad O \quad CH_3 \\ N \qquad \| \quad | \\ H_3C \diagup \diagdown CH_2-C-O-CH-CH_3 \\ Cl^- \qquad\qquad DAMP \end{array}$$

28.6 ppm

$$-CH_2-CH-\!\!-\!\!CH-\overset{\bullet}{C}H_2-\overset{\bullet}{C}H_2-CH-\!\!-\!\!CH-CH_2-$$
$$H_2C\diagdown \overset{+}{N}\diagup CH_2 \qquad H_2C\diagdown N\diagup CH_2$$
$$H_3C \diagup \diagdown CH_3 \qquad H_3C \diagup \diagdown CH_3$$
$$Cl^-$$

In the pairs *N*-vinylcaprolactam–DMAA and VP–DAMP this phenomenon is not realized ($r_{DMAA} = 3.42$, $r_{VCL} = 0.88$ and $r_{DAMP} = 4.39$, $r_{VP} = 0.25$). In these systems the copolymers formed are enriched by chainlinks of diallylammonium monomers at all compositions of an initial mixture of the monomers [123].

The distribution of chainlinks of polymer chain obtained from the monomers at the conditions of azeotropic polymerization was displayed by the ^{13}C NMR technique [123].

In order to estimate chainlink distribution, the following were used as analytic signals: the multiplet of chemical shifts (from 47.1 to 48.7 ppm) of carbon atom in $-C_xH-$ chain group in VP blocks and an unresolved doublet signal at

48.3–47.1 ppm

$-CH_2-CH-CH_2-\overset{\bullet}{C}H-CH_2-CH-$

N—C=O   N—C=O   N—C=O

180.3 ppm

δ

32.6 ppm   26.9 ppm

$-CH_2-CH-CH-\overset{\bullet}{C}H_2-CH_2-CH-\overset{\bullet}{C}H_2-CH-CH-CH_2-$

72.8 H_2C + CH_2 N—C=O H_2C CH_2
         N   N
    H_3C CH_3 H_3C CH_3
      Cl⁻

28.6 ppm of the CH_2-CH_2- bridge group connecting together five-membered rings of diallylammonium monomer.

There are two additional signals in the spectra of copolymers with chemical shifts at 26.9 and 32.6 ppm together with a signal at 28.6 ppm. The signal at 26.9 ppm is referred to $-CH_2-$ of a bridge group of a DMAA chainlink neighbouring VP $-CH_2-$ chain. The signal at 32.6 ppm is attributed to $-CH_2-$ of the bridge group of the DMAA link nearest neighbour to the methine group in PVP. The signals, especially that at 32.6 ppm, are broad and unresolved due to the presence of various triad and pentad sequences in copolymer chains [123].

It should be noted that the intensity changes in a range from 47.1 to 48.3 ppm. At the decrease of VP contents in copolymers the reduction of total intensity of the multiplet relative to that of the signal at 45.0 ppm is observed, together with the intensity redistribution of signals. The increase of signal intensity at 48.3 ppm at the appropriate reduction of that at 47.1 ppm indicates the increase of the contents of isotactic triads in comparison with syndiotactic ones. New additional signals were found in the spectra of copolymers in the fields 48.5–50.5 ppm, indicating the complication of configurational sequences involving $-C_2H-$ carbon atom of PVP.

The formation of alternating fragments in copolymers is testified by the change of a signal of a carbon atom in C=O of a pyrrolidone ring. The chemical shift of this carbon is quite sensitive to the close arranged atoms. In fact, a wide, slightly split signal at 180.3 ppm is observed for pure PVP, becoming narrower at 181 ppm with the increase of DMAA content in copolymers. In this case, a fraction of 'isolated' C=O groups of the copolymer chain grows together with an enhancement of their signal intensity.

Considering the intensity ratios of signals at 181.0 and 180.3 ppm (Table 2.5) it is possible to estimate the fraction of VP and DMAA block fragments in the structure of copolymers, and also that of alternating fragments of these two monomers.

Table 2.5   The intensity ratios of signals at 181.0 and at 180.3 ppm and that of signals at 48.5–50.5 and at 48.3–47.1 ppm in the ^{13}C NMR spectra of copolymers [123]. Reproduced with permission of Nauka

Copolymer composition DMAA : VP (mol %)	Intensity ratio of peaks	
	At 181.0/at 180.3 ppm	At 48.5–50.5/at 48.3–47.1 ppm
10 : 90	0.19	0.23
20 : 80	0.35	0.27
30 : 70	0.80	0.90
40 : 60	0.68	0.70
50 : 50	0.38	—
70 : 30	0.25	—

Maximum values of the ratio of peak intensity are found in a range from 30 to 50% of diallylammonium monomer in copolymers. It means the presence of a very large number of alternating pairs VP + DMAA (Table 2.5).

N-VINYLPYRROLIDONE AND N-VINYLFORMAMIDE

Pairs of vinyl monomers, possessing the reactivity ratios equal to one, are an interesting object of water-soluble polymer chemistry. The equal reactivity of monomers enables one to synthesize copolymers of tailored composition and high yield. The study of such monomer systems favours our understanding of the factors of their radical copolymerization and can help in finding new monomer pairs of similar reactivity ratios.

In the case of VP and VF it turns out that the copolymer composition corresponds to the mole ratio of the monomer (VP + VF) in the feed, at copolymerization reaction of VP with VF conducted at 65 °C in isopropanol solutions (to 7–10% conversion). Both methods (IR spectroscopy and ^{13}C NMR) give almost equal (within the experimental error of 5–10%) compositions of the copolymer [125]. Thus, this pair of *N*-vinylamides (VP and VF) enter so-called azeotropic copolymerization ($k_{11} = k_{22} = k_{12} = k_{21}$ and $r_1 = r_2 = 1$). Such behaviour of the two monomers in copolymerization is due to the structure of amide groups in VP and VF:

$$\begin{array}{cc}
CH_2{=}CH & CH_2{=}CH \\
\quad\backslash & \quad\backslash \\
\quad N{-}C{\nwarrow}^{O} & \quad N{-}C{\nearrow}^{H} \\
\quad\diagup \quad \backslash & \quad\diagup \quad \backslash \\
H \qquad H & H \qquad O
\end{array}$$

70% of *trans*-form   30% of *cis*-form

Equal reactivity of VP and VF is probably related to specific features of the conformational structure of the formamide side group, dipole–dipole interactions and hydrogen bonds which could arise in $(VF)_n$ associates of monomer molecules in solution.

It is noteworthy that in the PVF homopolymer and all copolymers, from 10 to 90% of the VF, the side amide group exists as a *trans*-conformer (see section 2.4) [125]:

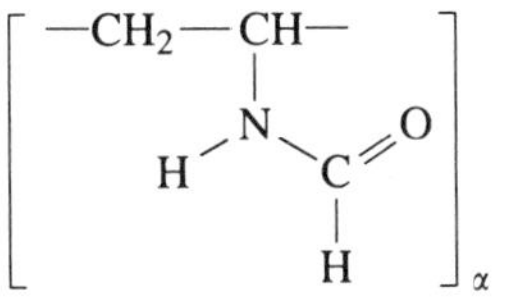

$$
\left[
\begin{array}{c}
-CH_2-CH- \\
| \\
H^{\diagdown}N^{\diagdown}C^{\diagup O} \\
| \\
H
\end{array}
\right]_\alpha
$$

The fact that almost all groups in PVF and VP + VF copolymer chains are in this conformer suggests that, during the chain propagation, favourable conditions for the association of VF molecules in the *trans*-form occur near the terminal radical. The units nearest to the terminal radical produce longer sequences of hydrogen bonds between amide groups of the chain and monomer molecules than in the case of a solution containing only monomer.

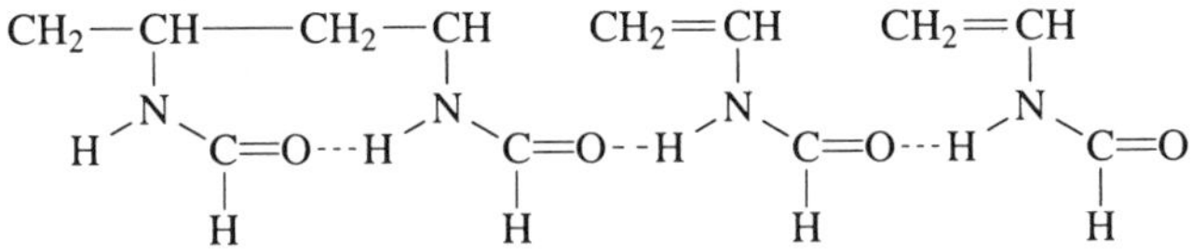

$$
CH_2-CH \text{——} CH_2-CH \quad CH_2=CH \quad CH_2=CH
$$

As the VP molecule is in *trans*-form, it can join the chain by hydrogen bonding between the C=O group of VP and the NH group of VF.

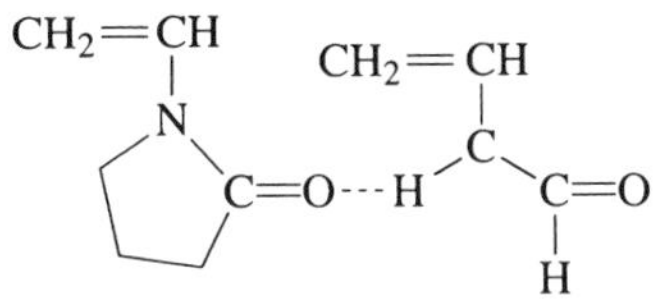

$$
CH_2=CH \qquad CH_2=CH
$$

VP molecules could also be involved in dipole–dipole interactions with the C=O group of VF at different orientation of the latter. The heat of association of molecules in *trans-trans*-form due to hydrogen bonding is apparently similar to the heat of association of VP and VF due to dipole–dipole interactions [125].

N-VINYLPYRROLIDONE AND VINYLPYRIDINE

The important problem of the synthesis of medicinal copolymers based on VP is not only the maintenancce of MW regulation but also the regulation of compositional distribution. At the same time VP low reactivity at radical copolymerization complicates the preparation of water-soluble copolymer of a certain compositional distribution. One of the solutions of this problem is to maintain distribution during the synthesis process [126].

Table 2.6   Reactivity ratios in copolymerization of VP (r_1) with vinylpyridines (r_2) [126]

Pyridines	r_1	r_2
4-vinylpyridine	0.0097 ± 0.0015	9.8 ± 1.5
2-vinylpyridine	0.014 ± 0.002	12.4 ± 2.3
2-methyl-5-vinylpyridine	0.036 ± 0.006	13.0 ± 2.0

The radical polymerization of monomers of very different reactivities at constant ratio of concentrations was realized on examples of 4-vinylpyridine (4-VP), 2-vinylpyridine (2-VP) and 2-vinyl-5-methylpyridine (MVP) in ref. [126].

Fedorov and co-workers [126] solved this problem, applying periodical analysis of the composition of reaction mixture by means of high-performance liquid chromatography (HPLC) and adding a more active monomer at a rate, calculated on the basis of this analysis.

The results of the work (Table 2.6) testify that reactivity of vinylpyridines at radical copolymerization with VP is very high (reactivity ratios differ by hundreds of times) and decreases in the order 4-VP > 2-VP > MVP, thus corresponding to theoretical views on the effect of substituents on the reactivity of double bond and stability of the corresponding radicals. In addition, a good fit of experimental data and calculated curves indicates that a classic Mayo–Lewis scheme is applicable to the systems under investigation.

Also, it was shown that the application of the Fineman–Ross method of reactivity ratios calculation is in this case undesirable. The calculations made under this approach are extremely unreliable for monomer pairs of very different reactivities. Instead the calculations were performed using minimization by the Nedler–Mead simplex method and the Kelen–Tudos method of linearization, and obtained data on VP and vinylpyridine reactivity was found identical [126].

It is also worth noting that performance of the synthesis at constant ratio of monomer concentration permits one to prepare copolymers of narrow composition distribution.

N-VINYLPYRROLIDONE AND METHACRYLIC ACID

The reactivity of *N*-vinylamides at radical polymerization largely depends on the nature of the solvent, in particular, of the protic solvent. This peculiarity of behaviour of these monomers is displayed at the copolymerization reaction, e.g. between VP and methacrylic acid anion (MAA) [127]. The change of VP hydrated surrounding by monomer concentration variation in water (5 and 50 wt%) is reflected on VP and MAA reactivity at copolymerization reaction (see Table 2.7).

A marked increase of VP reactivity relative to MAA is visible at going from concentrated solution (50 wt%) to a diluted one (5 wt%). The VP monomer becomes even more active in comparison with MAA in a dilute solution.

Table 2.7 Reactivity ratio for VP and MAA in aqueous solutions with various monomer concentrations [127]. Reproduced by permission of Nauka

Total monomer concentration (wt%)	r_{VP}	r_{MAA}
5	1.07 ± 0.02	0.55 ± 0.03
50	0.1 ± 0.01	4 ± 0.14

Note: Conditions of a polymerization: 65 °C, pH 8.0, [AIBN] 2×10^{-3}, NH_3

N-VINYLPYRROLIDONE AND ACRYLIC ACID

A solvent causes also a strong influence on VP copolymerization with acrylic acid [128–131]. As seen in Table 2.8, monomer reactivity of acrylic acid in polar environment (DMF or pure monomer) at copolymerization reaction is much higher than that of VP.

VP reactivity rises relative to that of acrylic acid at the performance of copolymerization reaction in protic solvent (Table 2.9). In fact, at going from DMF (Table 2.8) to water–alcohol mixtures (Table 2.9) the value of r_{VP} grows markedly from 0.015 up to 0.06–0.17 with the decrease of that of acrylic acid from 0.5 to 0.14. These findings are of interest in practice for the purpose of copolymer preparation from VP and acrylic acid with high yield and known composition.

However, at the performance of copolymerization reaction of VP and acrylic acid it is necessary to take into account a side reaction of VP hydrolysis occurring at pH < 6.0 and resulting in a reduction of monomer concentration [126].

Table 2.8 Monomer reactivity ratio of VP with acrylic acid in DMF and block

Experimental conditions	r_{AA}	r_{VP}	Reference
In block (75 °C)	1.3 ± 0.2	0.15 ± 0.1	[2]
In block (20 °C)	0.48 ± 0.04	0.05 ± 0.01	[130]
DMF (20 °C)	0.67 ± 0.05	0.03 ± 0.01	[130]
DMF (60 °C)	0.50 ± 0.05	0.015 ± 0.005	[131]

Table 2.9 Monomer reactivity ratio of VP and AA in water-alcohol mixture [128]. Reproduced by permission of Nauka

Solvent	r_{vp}	r_{AA}	Calculation approach
Water–methanol	$0.046 + 0.002$	$0.189 + 0.01$	Kelen–Tudos
Water–isopropanol	$0.137 + 0.007$	$0.143 + 0.014$	Kelen–Tudos
Water–tert-butanol	0.170 ± 0.008	0.27 ± 0.01	Kelen–Tudos

N-VINYLPYRROLIDONE AND CROTONIC ACID

VP copolymerization with crotonic acid (CA) is performed by γ-ray irradiation. Solovsky and co-workers [132] established that the copolymerization reaction between VP and CA in ethanol (or isopropanol), containing 85 mol% of VP and 15 mol% of crotonic acid, proceeds with a high yield. At radiation dose of 45×10^4 Gy the yield of copolymer reaches 93–95% and independent from concentration of monomers in an initial mixture in a range from 1 to 5 mol/l and from the nature of solvent. Monomer reactivity ratios of VP and CA was obtained by various initiation methods (at radiation-induced polymerization $r_{VP} = 1.00 \pm 0.05$ and $r_{CA} = 0.01 \pm 0.01$; at polymerization with free radical initiator $r_{VP} = 0.85 + 0.05$ and $r_{CA} = 0.02 + 0.02$) are virtually the same [132].

At a constant dose of γ-rays from a ^{60}Co source the increase of CA contents in an initial monomer mixture reduces conversion per cent and decreases contents of the acid in copolymer.

This phenomenon is caused by an increase of probability of chain termination reaction with the participation of inactive monomer (CA), inclined to degradation chain transfer. It is necessary to note that the yield of copolymers, containing 10–20 mol% acid groups, is rather high (86–93 wt%) [132]. Monomer concentration in ethanol regulates MW distribution, narrowing it from 3.0 to 1.4 at [monomer] = 3.0 and 0.3 mol/l respectively.

As this copolymer of narrow MW distribution is of interest at the development of new biologically active compositions for medicine, the radiation

Table 2.10 Reactivity ratios for *N*-vinylpyrrolidone (VP), *N*-vinylcaprolactam (VCL), *N*-vinylacetamide (VAA) or *N*-vinyl-*N*-methylacetamide (VMA) (r_1) with vinylacetate (VA) in copolymerization reactions

Monomer	r_1	r_{VA}	Reaction conditions	Measurement method	Ref.
VP	3.3	0.205	In block	Nitrogen content	[133]
	2.28 ± 0.19	0.237 ± 0.037	In block	Nitrogen content	[134]
	2.7	0.19	In block	Nitrogen content	[135]
	2.8 ± 0.1	0.25 ± 0.02	In block	Nitrogen content	[136]
	2.3	0.26	Dioxan	GLC	[137]
	3.108	0.348	In block	^{1}H NMR	[138]
VCL	0.63	0.31	In block	Nitrogen content	[139]
	2.2 ± 0.2	0.32 ± 0.04	Ethanol	IR-spectroscopy	[140]
	1.33 ± 0.1	0.3 ± 0.1	Butanol	^{1}H NMR	[121]
VAA	5.5	0.6	Methanol	^{1}H NMR	[39]
	2.1	∼ 0	Benzene	^{1}H NMR	[39]
VMA	4.79	0.57	Dioxan	GLC	[145]

Table 2.11 The reactivity ratios for N-vinylpyrrolidone (VP) with various co-monomers

No.	Co-monomer	r_{VP}	r_2	Reaction conditions	Ref.
1	N-vinylphthalimide	0.35 ± 0.02	1.28 ± 0.04	Dichloroethane (65 °C), AIBN	[141]
2	Diethylacetalacrolein	3.02 ± 0.02	0.01 ± 0.01	In block, (65 °C), AIBN	[142]
3	Methacrylacetone	0.02 ± 0.005	5.38 ± 0.05	In block (65 °C), AIBN	[143]
4	N,N-dimethylamino ethyl methacrylate	0.21	0.37	In block, (65 °C) AIBN	[139]
5	Iodide salt triethylammonium- ethyl methacrylate	1.69	0.33	In block (65 °C) AIBN	[144]

Table 2.12 The reactivity ratios for N-vinylacetamide (VAA) [39], N-methylacetamide (VMA) [145] or N-vinylformamide (VF) [146] (M_1) with various co-monomers (M_2)

N-vinylamide	Co-monomer	Reaction conditions	r_1	r_2
VAA	Acrylamide	Ethanol (60 °C) AIBN (1%) Fineman–Ross method	0.3	1.4
VF	Acrylamide	Water (50 °C), VA-44, Fineman–Ross method,	0.053 ± 0.023	0.534 ± 0.111
VF	Sodium acrylate	Water (45 °C),VA-44	0.29 ± 0.11	0.65 ± 0.03
VAA	Methyl methacrylate	Ethanol	0.19	2.65
VAA	Methyl methacrylate	DMF (60 °C), AIBN	0.01	2.1
VF	Butyl acrylate	THF (65 °C), AIBN Dioxan (60 °C)	0.061 ± 0.02	0.54 ± 0.09
VMA	Styrene	AIBN	0.01	19.25

Note: VA-44 is 2,2′-azo bis (N,N' dimethyleneisobutylamidine)dihydrochloride.

method of initiation recommnended by Solovsky, *et al.* [132] is quite promising in practice.

Concerning the data on copolymerization constants of N-vinylamides with other monomers, this is represented in three tables (Tables 2.10, 2.11, 2.12) and could be used as a good reference source [133–146]. VP, VCL, VAA and VMA reactivities to vinylacetate (VA) in copolymerization reaction are demonstrated in Table 2.10, and the reactivity ratios for VP, VAA, VF and VMA with various co-monomers are shown in Tables 2.11 and 2.12.

2.8 THE MOLECULAR WEIGHT CHARCTERISTICS OF POLYMERS

MOLECULAR WEIGHT OF POLYMERS AND ITS DETERMINATION

Molecular weight (MW) and molecular weight distribution of poly-*N*-vinyl-amides are the main physicochemical parameters determining the area of their application [2, 20, 43, 70, 76]. It concerns, in particular, PVP which is widely used in medicine and in industry. The properties of PVP materials vary in accordance with the average MW.

PVP is marketed for its various uses at different average molecular weights ranging from 2500 to 1 200 000. The different molecular weight materials are distinguished by *K*-number. The *K*-value is usually determined at 1% wt/vol of a given PVP sample in aqueous solution. The relative viscosity is obtained with an Ostwald capillary viscometer and the *K*-value is derived from Fikentscher's equation [67, 111]. So, PVP *K*-12, *K*-15, *K*-17, and PVP *K*-30 are included into both human and veterinary injection preparations. PVP *K*-25 and *K*-30 are used in the pharmaceutical, cosmetic and food industries.

Therefore, PVP products, particularly medicinal, are characterized either by MW value or by an experimental *K*-number (Table 2.13). Low-molecular-weight PVP of $M_w \approx 10\,000$ (12 600 ± 2600 or 8000 ± 2000) is manufactured for Russian medical use. PVP pharmaceutical grades are marketed under the trade marks Kollidon (BASF, Germany) and Plasdone (ISP, USA).

For a long period of PVP application (40–50 years) special attention of researchers was paid to MW determination.

Measurements of PVP average M_w were performed using sedimentation [147, 148], light scattering [109, 149–152, 154], gel-permeation chromatography (GPC) [153] and other methods. This attention to MW determination is caused by important practical requirements. Therefore, the use of PVP in medicine as a basis of a large number of medicinal preparations for various purposes requires PVP manufacturers to adopt strict conditions because a polymer of the precise required MW has to be produced and a quick MW analysis has to be performed on industrial production of PVP.

Table 2.13 PVP *K*-number and average MW ($\overline{M}_w$) of PVP for pharmaceutical industry [111]

K-value	Average $\overline{M}_w$
12 (11–14)	3900 (3100–5700)
17 (16–18)	9200 (7900–10 800)
25 (24–27)	26 000 (23 000–51 000)
30 (28–32)	42 000 (35 000–51 000)
90 (85–95)	1100 000 (90 000–1300 000)

Table 2.14 The values of K and α in the equation: $[\eta] = KM_w^{\alpha}$

K $\times 10^4$	α	Measurement conditions	Dispersity	Range of M_w	Year of report	Ref.
1.4	0.7	25 °C, H_2O	Fractional	$10^4 - 2 \times 10^5$	1951	[147]
3.38	0.63	30 °C, H_2O	Fractional	$10^4 - 2 \times 10^5$	1951	[113]
0.41	0.85	25 °C, H_2O	Fractional	$10^4 - 4 \times 10^4$	1653	[149]
1.9	0.68	25 °C, H_2O	Fractional	$2 \times 10^4 - 1 \times 10^6$	1953	[150]
1.3	0.68	25 °C, H_2O	Non-fractional	$2 \times 10^4 - 1 \times 10^6$	1953	[150]
5.65	0.55	25 °C, H_2O	Non-fractional	$1 \times 10^4 - 3 \times 10^5$	1958	[151]
6.76	0.55	25 °C, H_2O	Fractional	$10^4 - 3 \times 10^5$	1958	[151]
1.7	0.67	25 °C, H_2O	Non-fractional	$10^4 - 10^5$	1957	[152]
2.2	0.65	25 °C, H_2O	Non-fractional	$5 \times 10^3 - 5 \times 10^5$	1980	[153]
3.1	0.61	25 °C, H_2O	Non-fractional	$5 \times 10^3 - 10^6$	1984	[149]
2.32	0.65	25 °C, 50/50 v/v H_2O + methanol with 0.1 N $NaNO_3$	Non-fractional	$5 \times 10^3 - 10^6$	1987	[155]

The viscosity method meets these requirements, as one needs a short period of time to measure intrinsic viscosity ($[\eta]$) of the polymers obtained. The value of $[\eta]$ is usually determined by measuring the relative viscosity in a series of concentrations and extrapolating η_{SP}/C to zero.

The empirical Mark–Houwink equation links the value of $[\eta]$ and MW of polymers, i.e. $[\eta] = KM^{\alpha}$, where K and α are empirical constants, dependent on the conformational state of a macromolecule in solution and on the character of polymer interaction with solvent respectively and M is the average molecular weight (M_{η}, or M_w or M_n) measured by an independent physical method.

Since 1951 a large number of researchers have made efforts to establish the values of K and α for PVP. These values obtained by various authors are given in Table 2.14.

Possible reasons of such a variety of the values of K and α are stated in refs [150, 151, 155]. These reasons could be various contents of impurity in commercial samples, diverse ways of synthesis, resulting in polymer preparation with various contents of low-molecular-weight fractions at identical characteristic viscosity and the improvement of measuring techniques.

MW control at PVP fabrication until now has been performed using the Mark–Houwink equation ($[\eta] = KM_w^{\alpha}$), where $K = 1.4 \times 10^{-4}$ and $\alpha = 0.7$ (ISP, USA) [111], $K = 3.1 \times 10^{-4}$ and $\alpha = 0.61$ (BASF, Germany) [154], $K = 1.4 \times 10^{-4}$ and $\alpha = 0.7$ for PVP of $M_w = 12\,600 \pm 2600$ (Russia) and $K = 2.2 \times 10^{-4}$ and $\alpha = 0.65$ for PVP of $M_w = 8000 \pm 2000$ [109, 153].

The values of $\alpha = 0.6$–0.7 indicate the fact that water is a good solvent of PVP. The measurement data of K and α for PVCL and PVMA are shown in Table 2.15.

Table 2.15   Values of K and α for poly-*N*-methylacetamide (PVMA), poly-*N*-vinylpiperidone (PVPI) and poly-*N*-vinylcaprolactam (PVCL) (light scattering technique)

Polymer	$K \times 10^4$	α	Solvent	Range of MW	Ref.
PVMA	0.73	0.75	H_2O	10^4–10^6	[88]
PVPI	3.5	0.59	H_2O (20 °C)	5×10^4–4×10^5	[86]
PVCL	3.5	0.57	H_2O (25 °C)	5×10^3–10^6	[156]
PVCL	1.8	0.65	Methanol	5×10^3–10^6	[156]
PVCL	1.5	0.68	Ethanol	5×10^3–10^6	[156]
PVCL	1.3	0.70	Methanol	5×10^3–10^6	[156]
PVCL	1.1	0.69	H_2O, 20 °C	4×10^4–10^6	[86]

MOLECULAR WEIGHT DISTRIBUTION OF PVP

Knowledge of MW distribution of PVP obtained by radical polymerization is of major importance as the polymer is included in a wide range of medicinal preparations [41, 111]. So, PVP of MW = 8000 ± 2000 is used in 'Haemodesum-N' preparation, and PVP of MW = 12 600 ± 2 700 is a base of 'Hemodezium' preparation. These intravenous injection preparations represent water–salt solutions involving 6% of PVP [157], widely used in Russia as blood substitutes in detoxication. PVP of average MW (30 000 ± 5 000) underlies intramuscular injection preparations for prolonged action, such as Insipidin retard [158], 'Morphilongum' [43] and others.

Numerous studies concerning excretion and metabolism of PVP after intravenous administration to animals and man [70, 111, 158–161] established that the rate and the extent of clearance of PVP macromolecules through kidneys are dependent on molecular size. The range of PVP molecular weight in urine is often lower than that of the administered material, showing selective excretion. Biochemical studies revealed that healthy people can excrete PVP molecules of a molecular radius as high as 4 nm, which corresponds to molecular weight of around 7×10^4. A glomerulus in a kidney is highly permeable to molecules of molecular weight less than 3×10^4, but is relatively impermeable to molecules of MW $> 7 \times 10^4$. As polymer material of large molecular weight is not excreted by a kidney, significant retention in tissues may occur. So, PVP of MW $< 3 \times 10^4$ is quickly excreted within several (3–4) days, as measured by radioactive procedure [68, 106]. Macromolecules of MW between 3×10^4 and 1×10^5 are eliminated during several months. Macromolecules of MW $> 1.1 \times 10^5$ are retained in body for several years and could cause diseases due to excessive storage of PVP [111].

A study of molecular weight distribution of PVP was carried out using gel permeation chromatography (GPC) with dextran gels [109, 110, 162] or cross-linked agarose [163].

So, Sepharose CL-4B, Sephadex G-75 and Sephadex G-25, packed in a column, with their heights ratio being 3.5 : 1 : 1.5, separate PVP samples of MW in a range from 10^3 up to 5×10^5 in 0.9% NaCl aqueous solution as a mobile phase. Calibration dependence of MW on elution volume of a column is found using PVP samples of known MW (M_w by light scattering technique, M_n by ebullioscopy) [109, 110]. This approach is useful to control fractional composition of PVP samples for medicinal purposes, i.e. PVP of M_w in a range from 3×10^3 up to 4×10^4.

MW distribution in samples of PVP K-18 and PVP K-14.2 was determined by GPC [156], displaying contents of various fractions of MW ranging from 500 to 5×10^4. For example, K-18 and K-14.2 samples include 10% and 24% of PVP fractions of $M_n < 10^3$ and 2% and 0.05% of fractions of $M_n > 3 \times 10^4$.

In order to apply high-performance liquid chromatography using the Waters Model GPC, the search for a stationary phase, not interacting with PVP macromolecules, was conducted [159]. The linear dependence of log ($[\eta] \times M$) on the elution volume of PVP and polyethyleneoxide (PO) standards in mobile phase of 50 : 50 (v/v) CH_3OH/H_2O containing 0.1 M $LiNO_3$ was established for a semi-rigid polymer gel containing $-CH_2-CHOH-CH_2-O-$ groups (TSK-Gel type-PW, Toya Soda Co., Japan) [155, 164].

This calibrating dependence was used in the calculation of MW distribution of PVP samples manufactured by ISP (USA) (K-90, K-60, K-30, K-15) [155]. Exact values of M_w of PVP samples were also determined using high-performance liquid chromatography (HPLC) with application of low angle laser light scattering (LALLS) detector. Unfortunately, LALLS is incapable of estimating precisely the portion of low molecular weight fractions that result in overvalued average number MW (M_n) and, accordingly, in the reduction of D = $\bar{M}_w/\bar{M}_n$.

At the same time, PVP MW distribution calculation using universal calibration is complicated by the chromatographic peak breadth of fractions output, in particular those of low MW. In this case the value of $\bar{M}_n$ become undervalued, with $\bar{M}_w/\bar{M}_n$ being overvalued [150, 159].

The data in refs [155, 164] indicate the fact that different values of $\bar{M}_w/\bar{M}_n$ and also of K, namely from 2.32×10^{-4} to 0.92×10^{-4}, and α, from 0.65 to 0.7, in the Mark-Houwink equation (solvent 50 : 50 (v/v) CH_3OH/H_2O with 0.1 N $LiNO_3$) can be calculated investigating PVP commercial samples by various techniques.

One can assume that variation of K- and α-values in this equation, observed by different authors during a long period of time (Table 2.14), is caused just by inaccurate determination of the contents of low-molecular-weight fractions. Meanwhile, the contents of these fractions can vary considerably from batch to batch of polymer commercial lots, causing fluctuations in the determination of $\bar{M}_w/\bar{M}_n$. It seems to be due to the fact that the traditional process of VP polymerization in a system of VP + H_2O_2 + NH_3 + H_2O is difficult to control. Various factors affect VP polymerization reaction (see section 2.5), being still unclarified.

Correct estimation of the value of α in the equation ($[\eta] = KM^{\alpha}$) characterizing the interaction of PVP macromolecules with a solvent (water) and conformational state of macromolecules in a solution was performed with the application of low MW polymer samples prepared in isopropanol as chain transfer and at various VP concentrations, ranging from 0.8 to 15 vol% (10–20% conversion) in the presence of AIBN [116].

A series of polymers of MW ranging from 2×10^3 up to 7×10^4 was prepared. The factors affecting the contents of low-molecular-weight fractions in PVP samples of various MW are excluded in this case. Dependence of log $[\eta]$ on log ($\bar{M}_n$) is approximated by the Mark–Houwink to be $[\eta] = (3.0 \pm 0.35) \times 10^{-4} \cdot \bar{M}_n^{0.65 \pm 0.17}$, i.e. $\alpha = 0.65 \pm 0.017$ for pure water at 25 °C. The average number molecular weight ($\bar{M}_n$) is measured by ^{1}H and ^{13}C NMR methods. High resolution of proton signals of terminal groups and of the main chain in the ^{1}H NMR spectrum has allowed the estimation of $\bar{M}_n$ from the areas ratio of the appropriate signals.

GPC method of PVP molecular weight distribution estimation, with the application of TSK-GEL columns [155, 164], has recently become standard. Researchers apply different mobile phases, namely a mixture of methanol and water (50 : 50 v/v) involving $LiNO_3$ (0.1 M) [155], an aqueous solution of sodium acetate [94,148] or water–methanol solution of 80% methanol content [165].

This method allows one to determine the MW of PVP samples in a range of high MWs. For example, MW distribution of PVP samples of high molecular weight synthesized by irradiation-induced polymerization (γ-ray from a ^{60}Co source) in water and in alcohols, is characterized by the value of $\bar{M}_w/\bar{M}_n = 1.9$–$2.5$ in a range of $\bar{M}_w$ from 3×10^5 up to 1.5×10^6 [94].

3

Solvation (Hydration) and Conformational Transformations of Macromolecules in Solution

3.1 SOLUBILITY

The polymer solubility in that or any other liquid serves as primary information on interaction of this polymer with a given liquid. In a number of poly-N-vinylamides with common polar nature of the amide group, the structure of side substituents greatly affects the ability of polymer to dissolve.

WATER

Poly-N-vinylamides of aliphatic carboxylic acids, where at an atom of nitrogen and at the carbonyl group, act as substituents H, CH_3 and C_2H_5 at any combination, are well dissolved in water [106].

PVP dissolves in water at high concentration and precipitates only at 170 °C. PVCL dissolves in water at a temperature from 0 °C up to 30 °C; however, at 32 °C and above polymer precipitates in a separate phase [87, 166, 167]. The temperature of phase separation ($T_{ph.s.}$) or low critical solution temperature (LCST) varies with M_w of PVCL (Fig. 3.1).

The most appreciable increase of $T_{ph.s.}$ originates for PVCL macromolecules with $M_w < 50 \times 10^3$. At M_w decreasing from 2×10^6 to 50×10^3 the value of $T_{ph.s.}$ changes from 32.3 °C only to 35–36 °C, whereas in the M_w range from 50×10^3 to 6×10^3, $T_{ph.s.}$ grows significantly from 36 °C to 51 °C (Fig. 3.1) [63]. The concentration of polymer in water slightly affects the $T_{ph.s.}$ value.

Other poly-N-vinylamides, such as poly-N-vinyl-N-methylpropionamide, poly-N-vinyl-N-ethylpropionamide, poly-N-vinyl-N-propylacetamide are soluble in water up to $T_{ph.s.} = 80, 52$ and 48 °C respectively [106].

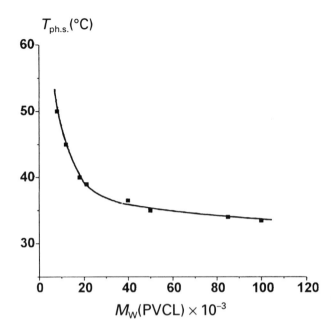

Fig. 3.1. Dependence of $T_{ph.s.}$ of PVCL in water on weight-average molecular weight (M_w) of macromolecules. [PVCL] = 0.4 wt% [63]. Reproduced by permission of Nauka.

It was found that phase separation of VCL copolymers in aqueous solution depends on the structure of the second monomer and the composition of the copolymer [121, 166].

Chainlinks of the second monomer, e.g. VP, being more hydrophilic than VCL, are responsible for the $T_{ph.s.}$ increase from 33 °C (pure PVCL) to 80 °C at 34 mol% of VP in copolymer [166]. The same effect on $T_{ph.s.}$ of PVCL exhibit chainlinks of *N*-vinyl-*N*-methylacetamide [121]. Hydrophobic chainlinks of vinylacetate (VA) decrease $T_{ph.s.}$ significantly: the copolymer containing 66 mol% VA precipitates at 5 °C (Fig. 3.2).

In the case of copolymers with OH groups this dependence has another characteristic (Fig. 3.2). It follows from Fig. 3.2 that with the increase of vinyl-alcohol contents in copolymer from 0 to 50 mol% the value of $T_{ph.s.}$ decreases from 33 °C (native PVCL) to 25 °C with a subsequent enhancement up to 52 °C for the copolymer with 87 mol% of vinylalcohol chainlinks. The minimum on a thermoprecipitation curve of VCL + vinylalcohol copolymer corresponds to the equimolar composition of the copolymer, which specifies the formation of intramolecular hydrogen bonds between OH and C=O groups, promoting intramolecular association. In IR spectra of films made from these copolymers

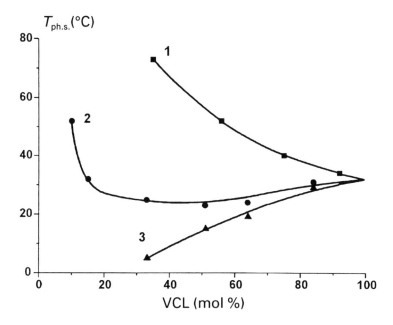

Fig. 3.2. Dependence of phase separation temperature of VCL copolymers with *N*-vinyl-*N*-methylacetamide (1), vinylalcohol (2) or vinylacetate (3) [121]. Reproduced by permission of Nauka.

$\nu_{C=O}$ of the caprolactam ring is shifted from 1630 cm^{-1} (15 mol% of vinyl-alcohol) to 1580 cm^{-1} (87 mol% of vinylalcohol chainlinks in the copolymer with VCL), testifying to the increase of a number of hydrogen bonds between chainlinks of various nature. Introduction of 'pure' polyvinylalcohol in a PVCL solution so that the composition of a mixture corresponds to copolymer composition does not, in practice, affect $T_{ph.s.}$

ALCOHOLS

Alcohols (methanol, ethanol, propanol, etc.) are good solvents for the majority of poly-*N*-vinylamides under investigation, apart from poly-*N*-vinyl-*N*-methyl-formamide, poly-*N*-vinylformamide and poly-*N*-vinylacetamide. In case of the last two polymers strong hydrogen bonds both inside coils and between different macromolecules prevent its dissolving in alcohols. For poly-*N*-vinyl-*N*-methylformamide, in which macromolecules have alternation of *cis*- and *trans*-conformers in side substituents (see Chapter 2), the strong intramolecular dipole–dipole interactions makes it insoluble even in methanol and other alcohols [106, 111].

AMIDES

Dimethylformamide, *N*-methylformamide, *N*-methylpyrrolidone and other amides are good solvents for most poly-*N*-vinylamides (PVP, PVCL, PVMA, etc.). It is necessary to note that in a series of investigated polymers poly-*N*-vinyl-*N*-methylformamide was found to be solvent only in amides with proton at a nitrogen atom [106, 111].

CHLORINATED HYDROCARBONS

Poly-*N*-vinylamides involving alkyl substituents at a nitrogen atom and at a carbon atom of the C=O group and PVP are dissolved in CH_2Cl_2 and $CHCl_3$. PVCL possesses the same property, also dissolving in CCl_4. Poly-*N*-vinyl-*N*-methylformamide and other amides involving hydrogen at a nitrogen atom are insoluble in these mediums.

AROMATIC HYDROCARBONS

Only PVCL has the ability to dissolve in toluene, benzene and xylene. For all other polymers considered they are precipitators.

HYDROCARBONS AND KETONES

Dissolving of poly-*N*-vinylamides in unsaturated hydrocarbons (heptane, hexane and others) and ketones (acetone) virtually does not occur.

Thus, the poly-*N*-vinylamides under consideration are characterized by a high capability to interact with proton-containing (water, alcohols) and polar molecules (e.g. DMF). The understanding of interaction peculiarities between macromolecules of this structure and solvent molecules, e.g. water, is of profound importance for detection of specific properties of these macromolecules in a complex formation with a large number of low-molecular-weight and high-molecular-weight compounds and interaction with cells of living organisms and other processes. In this connection knowledge of the physicochemical factors affecting these interactions will promote the location of water-soluble polymers with new properties valuable in practice.

3.2 HYDRATION OF MACROMOLECULES WITH AMIDE GROUPS

The establishment of solvent–macromolecule interactions, and in this case conformational transformations of a chain, and also properties of solvent molecules near to macromolecules, is an important problem in the physical chemistry of water-soluble polymers, in particular those which are close analogues of protein molecules due to the presence of amide bonds.

POLY-N-VINYLPYRROLIDONE-WATER (IR SPECTROSCOPY)

On an example of PVP using the IR spectroscopic method one can show the action of hydration of the polymer on a change of spectral characteristics of the $C=O$ group (the band due to the stretching vibrations of the carbonyl) and water molecules into a hydrate shell (the bands due to the stretching vibrations of D_2O) [120].

Significant changes of $v_{C=O}$, degree of asymmetry (g) and integral intensity ($S = \int D(v)d(v)$) of this band with the increase of the number of molecules per one chainlink (N) were found (Fig. 3.3)

The degree of asymmetry was calculated in the relation:

$$g = [(v_h - v_0) - (v_0 - v_1)]/(v_h - v_1) \times 100\%$$

where v_0 is the observable frequency of the band maximum, v_h and v_1 are frequencies of the band, measured at a level of its half-height with high-frequency (h) and low-frequency (l) sides of the band.

Introduction of one D_2O molecule per one chainlink of PVP (film) results in a significant lowering of $v_{C=O}$ from 1682 to 1660 cm^{-1}, transformation of the band form and increase of its integral intensity. These spectral changes are attributed to the formation of a hydrogen bond between $C=O \bullet\bullet\bullet D-O-D$.

A further increase of N up to five D_2O molecules lowers $v_{C=O}$ by 20 cm^{-1} and slightly enhances S (from 30 to 35 cm^{-1}). The band contour with increasing N becomes fourfold asymmetric and changes the asymmetry sign three times. Such dependence character of g on N testifies to the existence of four to five individual bands of $v_{C=O}$ with various spectral parameters.

It was found also that in a solution of polymer in D_2O ($N = 260$) the band of $v_{C=O}$ has a complex structure (two components with $v = 1640$ and 1660 cm^{-1}) [116]. Lowering of a solution temperature from 35 °C to 5 °C causes precise separation of the band into two components at $v_{C=O} = 1640$ and 1660 cm^{-1}.

In turn, amide groups of PVP affects, through hydrogen bond formation, the asymmetric (v_3) and symmetric (v_1) stretching vibrations of D_2O molecules co-operating with PVP chainlinks (Fig. 3.4). There is a decrease of these frequencies with increasing N from 0.2 to 3–5.

At $N > 2$ the decrease of the frequencies stops, indicating the presence of two D_2O molecules in close proximity to one carbonyl group. It also follows from plots on Fig. 3.4 that with reducing N the increase of frequencies prove the lower power of hydrogen bonds between the $D-O-D \bullet\bullet\bullet C=O$ groups in comparison with those between $D-O-D \bullet\bullet\bullet D-O-D$. In addition, it should be noted that the important fact is that in IR spectra of hydrated PVP there are no spectroscopic attributes, indicating the participation of unshared p-electrons of

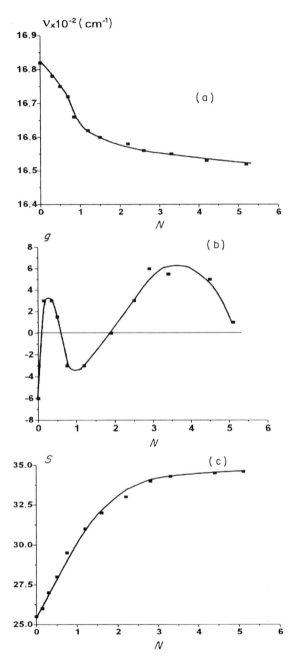

Fig. 3.3. Effect of hydration degree (N) on frequency (v_0) of the band maximum (a), para-meter of asymmetry (g) (b) and integral intensity (S) (c) of the C=O band in PVP [120]. Reproduced by permission of Nauka.

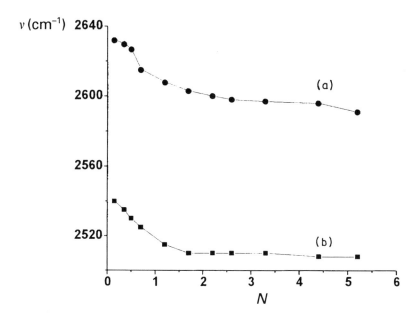

Fig. 3.4. Frequency change of the band maxima of symmetric (ν_1) (a) and asymmetric (ν_3) (b) stretching vibrations of D_2O depending on hydration degree (N) [120]. Reproduced by permission of Nauka.

the nitrogen atom in the hydration of PVP chainlinks [120]. Thus, in an aqueous solution the first hydrate layer near the chainlinks consists of two H_2O molecules. Two types of associates are observed in a dilute PVP solution: the first with one D_2O molecule ($\nu = 1660$ cm^{-1}) and the second with two D_2O molecules ($\nu = 1640$ cm^{-1}):

The lowering of temperature from 35 °C to 5 °C increases the fraction of associates with two molecules (II). As spectral parameters cease to change at $N = 5$ (Figs. 3.3 and 3.4), the second hydrate layer consists of three to four D_2O molecules. The action of water molecules of the second hydrate layer on $\nu_{C=O}$ is caused by the fact that the formation of hydrogen bonds between molecules of the second and the first layer increases the proton-donor ability of D_2O in complexes with C=O groups.

Considering the frequency change of C=O stretching vibrations in a hydrated PVP IR spectrum, one can estimate that the first two water molecules and then about three more molecules, connected with the previous ones, take place in direct proximity at the C=O group of a chainlink [116]. At the same time the common hydrate shell, surrounding the pyrrolidone chainlink as a unit, contains an appreciably greater number of water molecules than $N = 5$.

POLY-N-VINYLPYRROLIDONE AND WATER–ORGANIC MIXTURES (BY ^{13}C NMR)

The chemical shift of carbon (^{13}C) atoms in a carbonyl and in a PVP chain ($-C_xH-$) in a solution is defined not only by the contribution of electronic density on these atoms but also that of the solvent and spatial surroundings

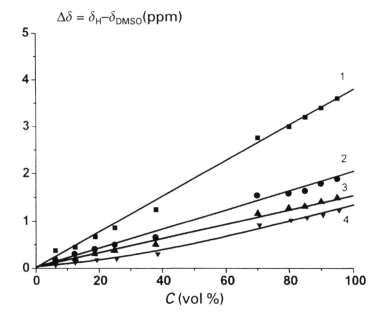

Fig. 3.5. Effect of mixture composition of DMSO and methanol on chemical shifts of carbon atoms in C=O (1) and in methine group of PVP chain (mm-(2), rr-(3), mr-(4)) [169]. Reproduced by permission of Nauka.

created by other fragments of a macromolecule [51, 168, 169]. In this case the change of chemical shift of the specified atoms is a reflection of more distant interactions arising from solvation (hydration) of chain sites located at the atoms.

In fact, the change in the nature of the PVP solvent is affected in the ^{13}C NMR spectrum of the polymer (Fig. 3.5 and Table 3.1). The displacement of chemical shifts (δ) of polymer groups in protic solvent in comparison with aprotic solvent (e.g. dimethylsulfoxide) displays the influence of solvation. As seen in Fig. 3.5, the displacement of a chemical shift of the carbon atom in the C=O group is greatest for water and decreases in a series of solvents: methanol > ethanol > n-butanol > chloroform [169]. It means that the C=O group forms the strongest hydrogen bonds with water molecules and the weakest hydrogen bonds with molecules of chloroform.

The solvation affects chemical shifts of other ^{13}C atoms of a chain and of a ring (Table 3.1). It should be noted that the displacement of the chemical shift of ^{13}C atoms of the chain, being in syndiotactic, isotactic and heterotactic triads, is different for all protic solvents, specifying different capabilities of the C=O groups in these triads to interact with the proton of the solvent molecule.

The shift of δ (C=O) in a downfield with decreasing concentration of PVP from 25 to 2.5 mol%, corresponding to an increase in number (N) of H_2O molecules per chainlink from 3 to 30, demonstrates the C=O hydration. The same shift of δ ($-C_xH-$) in iso-, hetero- and syndio-sequences indirectly indicates conformational transformations of PVP chains occurring with increase of the hydrate shell (Fig. 3.6). These atoms of the main chain are almost inaccessible to interaction with H_2O molecules, as shown by IR spectroscopy of PVP [120].

Table 3.1 Solvent nature effect on chemical shifts of ^{13}C PVP atoms in relation to those in DMSO at 50 °C [169]. Reproduced by permission of Nauka

Solvent	$\Delta\delta = \delta_0 - \delta_{DMSO}$ (ppm)						
	C=O	mm	mr	rr	C_4	C_2	C_3
H_2O	5.35	2.85	2.02	2.54	2.55	2.08	1.44
CH_3OH	3.93	2.14	1.38	1.50	1.72	1.49	1.25
C_2H_5OH	3.31	1.62	0.99	1.41	1.40	1.15	0.98
C_4H_9OH	2.55	1.35	0.72	1.14	1.12	0.94	0.86
$HCCl_3$	1.82	0.54	0.3	0.41	0.74	0.55	0.51

$$(-C_\beta H_2 - \underset{|}{C_x}H-)$$

$$H_2C\underset{|}{\overset{N}{\diagdown}}C=O$$
$$H_2C \underset{(2)}{\overset{(3)}{------}} CH_2$$

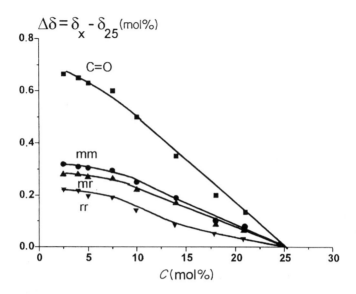

Fig. 3.6. The effect of PVP concentration in water on chemical (^{13}C) shifts of C=O and methylene groups [169]. Reproduced by permission of Nauka.

It is essential that these dependencies of δ (C=O) and δ (—C$_x$H—) on PVP concentration in water cease to change at 7.5 mol%, specifying the fact that each PVP chainlink is embedded in a shell of about 10–12 water molecules. At a more significant dilution, when $N > 10$–12, saturation of the hydrate shell ceases.

These data (IR spectroscopy and ^{13}C NMR) allow us to conclude that in close proximity to C=O some five H$_2$O molecules are likely to be present, whereas the remaining five to seven water molecules surround methylene groups of the ring at PVP concentration in water < 7 mol% at room temperature.

3.3 WATER IN POLYMER SOLUTION AND GEL

DIFFUSION OF WATER MOLECULES IN POLYMER SOLUTION

The introduction of PVP ($M_w = 30 \times 10^3$) or its low-molecular-weight analogue *N*-ethylpyrrolidone in water causes an appreciable lowering of the self-diffusion coefficient (D_{sd}) of water molecules, determined by a spin-echo technique [60], from 2.5×10^{-5} cm^2/s in bulk water to 0.5×10^{-5} cm^2/s in the presence of *N*-ethylpyrrolidone (20–25 mol%) and 0.25×10^{-5} cm^2/s in the presence of PVP at the same concentration.

Slowing down of mobility of the H_2O molecule occurs in an identical manner, both in the case of PVP and its analogue, up to $C = 10–12$ mol% of the added substance. Hence, the mobility of water molecules in the presence of N-ethylpyrrolidone or PVP decreases at the expense of interaction between H_2O molecules and $C\!=\!O$ groups of lactam rings.

As is established from viscosity measurements of mixtures of $H_2O + N$-ethylpyrrolidone [60], the greatest viscosity is observed at 25 mol% of N-ethylpyrrolidone. In other words, a complex is formed in the solution, consisting of three molecules of water per molecule of analogue.

At the same composition (analogue + H_2O) the value of D_{sd} ceases to fall, being 0.5×10^{-5} cm^2/s. In the case of the polymer + H_2O composition the change of D_{sd} ceases in the same range of concentration (25–30 mol%), testifying that about three water molecules are strongly connected to the $C\!=\!O$ group of the PVP chainlink. These data virtually do not differ from those obtained by IR spectroscopy. In addition, the results concerning the D_{sd} change against PVP concentration in water [60] and that of Fig. 3.6 (^{13}C NMR data [164]) show a certain correlation between the self-diffusion coefficient of water (D_{sd}) in a polymer solution and the number of hydrate molecules of water near to a PVP chainlink in the same solution in dependence on the mixture composition. In fact, only at [PVP] < 8–9 mol% (11–12 H_2O molecules per unit) the linear increase of water molecule mobility with decreasing polymer concentration is found, since free H_2O molecules appear in the solution increasing the value of D_{sd}. Similarly, after [PVP] < 7.5–8 mol% in water (about 12–13 molecules per unit) the increase of chemical shift of carbon (^{13}C) atom in $C\!=\!O$ stops, indicating complete saturation of the hydrate shell of chainlinks and appearance of free water molecules.

Thus, two methods testify to significant interaction of the PVP chainlink with H_2O molecules. It is necessary to allocate the first hydrate layer of two or three water molecules (IR spectroscopy [116], method of spin echo NMR [60]), the second hydrate layer involving two or three more molecules (total five or six) (IR spectroscopy [120]) and the third one containing additional five or six molecules (total 10–12 H_2O molecules per unit) (^{13}C NMR and method of a spin echo NMR at 25–30 °C) [60].

MELTING OF 'ICE WITH MACROMOLECULES'

The introduction of macromolecules in an aqueous solution exerts considerable action on structural organization of water associates, e.g. at a crystallization. The presence of such polymers affects the melting temperature of ice with macromolecules included, thermal effects of melting of ice-like crystal structures formed from water molecules at freezing of an aqueous polymer solution, and so on [170].

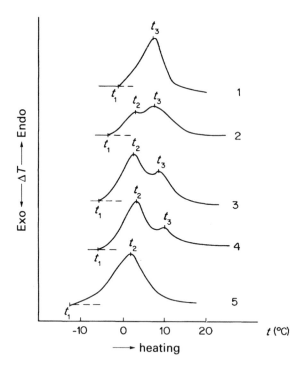

Fig. 3.7. Thermograms of heating of frozen water (1) and aqueous solutions of PVMA (2), PVMF (3), PVP (4) and PVCL (5), polymer concentration is 1 mol/1. Scan rate is 5 °C/min [170]. Reproduced by permission of Nauka.

The effect of the structure of poly-N-vinylamides used (PVP, PVCL, PVMF and PVMA having $M_w = 20 \times 10^3 - 30 \times 10^3$) on water crystallization is observed from the character of thermograms of melting of pure ice and frozen aqueous solutions (DSC method) [165]. In all cases a precise endothermic maximum near 0 °C (Fig. 3.7) takes place. For pure ice the maximum is unimodal, and heat of the endothermic effect is 333.9 J/g.

At a melting of frozen solutions of PVP, PVMA and PVMF a bimodality of thermal effect, being represented by two peaks (t_1 and t_2) on thermograms, is observed. The ratio of the peak heights is defined by the polymer concentration in water, and with growth of its contents the area of high-temperature peak decreases, with increasing the area of the low-temperature peak. At a concentration more than 2 mol/1 for all polymers investigated only one low-temperature peak is observed.

The presence of two peaks on the heating thermograms of frozen aqueous solutions of PVP, PVMF or PVMA is due to phase transition of the first kind

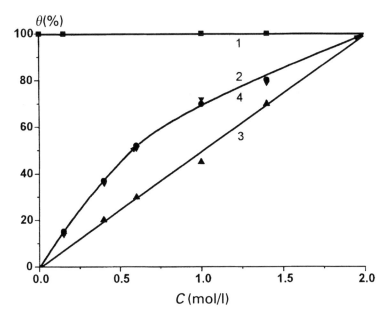

Fig. 3.8. Fraction of water 'interacting' with polymer (θ) in dependence to polymer concentration: PVCL (1), PVMA (2), PVMF (3) and PVP (4) [170]. Reproduced by permission of Nauka.

(melting) for two systems: (1) pure ice from water frozen without impurities in microscopic volumes; (2) ice-like formations with introduced macromolecules. In the latter case the macromolecules affect the structure formed at freezing of water 'interacting' with macromolecules, appreciably different from the crystal structure of pure ice.

The contents of the ice-like units and pure ice in frozen water–polymer solution depends on both the structure and concentration of dissolved poly-N-vinylamides (Fig. 3.8). Considering the areas of the peaks received by separation of doublet maxima, the ratio of indicated structural types of ice was estimated.

As follows from Fig. 3.8, significant deterioration within the ice structure appears with adding the polymer, and at $C_{pol} > 2$ mol/l for PVP, PVMA and PVMF the structure of ice-like formations differs from the pure ice structure, i.e. the macromolecules at such concentration effectively act on the whole water volume (Fig. 3.7). In this case, i.e. at 0 °C, each polymer chainlink is surrounded by an average of 20–22 molecules of water. The effect of the polymer structure is that at identical concentration (< 2 mol/l) the polymers fall into a certain order with increase of degree of structure disturbance within pure ice: PVMA < PVP < PVMF. Such a sequence indicates that from a range of poly-N-vinylamides considered, PVMA deforms the structural lattice of ice the least.

This fact can be interpreted considering the structural model of water with filling of cavities [171, 172]. The volume sizes of methyl group, included in PVMA, are close to those of cavities in an open-work lattice of ice. The methyl group, being fitted into hollow areas between molecules of water, should exert the least, in comparison with other substituents, destructive action on an ice lattice.

Simultaneously, macromolecules of PVCL in the greatest degree deforms the structure of ice-like formations. Therefore, an interesting peculiarity of thermograms of melting of frozen aqueous PVCL solutions [170] is that in all concentrations investigated (0.15, 0.5, 1.0, 1.5 and 2.0 mol/l) the thermograms have only one endothermic low-temperature peak (Fig. 3.7). In addition, the values of t_1 with increase of PVCL concentration are shifted to a lower temperature field.

These data mean that PVCL macromolecules effectively disrupt the structure of ice, even when if no less than the average of 500 H_2O molecules fall into one polymer chainlink. Probably, it is caused by complete stereodiscrepancy of volume caprolactam substituents in PVCL chain to cavities within a pure ice structure. In fact, the PVCL macromolecule, unlike other poly-*N*-vinylamides researched, has certain peculiarities of the chain microstructure and side residues (see Chapters 1 and 2). The microstructure of main PVCL chain consists of syndiotactic configuration sequences, and the caprolactam ring is rather voluminous and has specific 'chair'-type conformation.

The introduction of poly-*N*-vinylamides into water medium affects the enthalpy of melting of ice-like formations (Fig. 3.9). The value of ΔH_{mlt} is gradually reduced with increasing polymer concentration, from 334 J/g for ice in bulk water and up to 220–230 J/g at 2.0 mol/l, when polymer macromolecules attack all water molecules in a solution.

It is worth noting that the lowering of ΔH_{mlt} occurs in an identical manner for three polymers (PVP, PVCL and PVMF), to within experimental error (5%–10%) of the heat measurement, whereas in the case of PVMA the value of ΔH_{mlt} remains almost constant up to 1 mol/l. Only beyond this concentration does ΔH_{mlt} begin to reduce sharply and approach that of PVP and PVMF (226 J/g) (Fig. 3.9). This fact indicates the above-mentioned weak action of PVMA macromolecules on the structure of pure ice.

One should emphasize that in the case of PVP the melting enthalpy of ice-like formations does not virtually change in going from a diluted solution (0.15 mol/l of PVP) to a moderately concentrated one (up to 1.5 mol/l) ($\Delta H_{mlt} =$ 230–240 J/g, see data on Figs. 3.8 and 3.9). The calculated ΔH_{mlt} (230 J/g) coincides with that of a frozen aqueous PVP solution (2 mol/l), on the heating thermogram of which there is no peak of pure ice melting. This result testifies that each PVP chainlink in a range of polymer concentration from 0.15 up to 2 mol/l, in similar manner converts the structural organization within water associates of 20–22 molecules of water, making ice-like crystal formations that

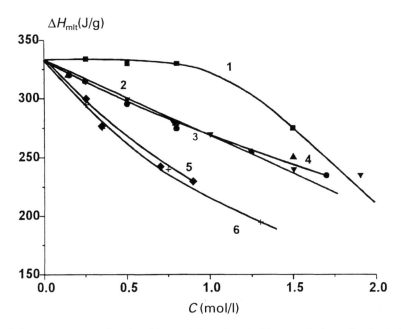

Fig. 3.9. Dependence of total melting enthalpy (ΔH_{mlt}) of frozen solutions of poly-*N*-vinyl-*N*-methylacetamide (1), poly-*N*-vinyl-*N*-methylformamide (2), PVP (3) PVCL (4), dimethylformamide (5), and dimethylacetamide (6) on its concentration in aqueous solution [170]. Reproduced by permission of Nauka.

require less thermal energy (230 J/g) for melting than in the case of pure ice. In addition, the constancy of ΔH_{mlt} of PVP-ice-like formations in a wide range of polymer concentration confirms the absence of 'non-freezing' water in this system.

A concentration effect of the polymers added on the melting temperature of 'ice-like formations with macromolecules' (Fig. 3.10) was observed [170]. As seen in Fig. 3.10, PVMA to a smaller degree, compared to PVP, PVMF and PVCL, reduces T_{mlt} of modified ice at all concentrations considered (from 0 to 2.0 mol/l). The character of T_{mlt} decreases with increasing polymer concentration and is almost identical for PVP, PVCL and PVMF. The value of T_{mlt} falls from 0 °C for bulk water to – 8 °C and – 12 °C when the polymer concentration is 1.0 and 2.0 mol/l respectively. It is noticeable that in the case of PVMA the values of T_{mlt} are higher than in the case of other polymers.

The low-molecular-weight analogues, namely dimethylformamide (DMF) and dimethylacetamide (DMA) render a stronger action on the H_2O structure. The values of ΔH_{mlt} and T_{mlt} decrease to a greater degree with increasing concen-

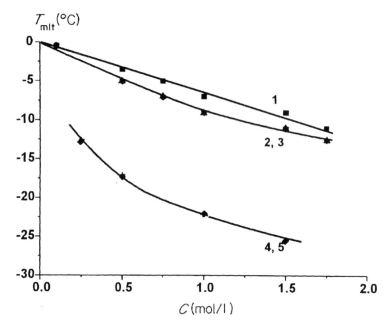

Fig. 3.10. The effect of polymer (PVMA-1, PVP-2 and PVCL-3) and low-molecular-weight analogue (DMA and DMF-4,5) concentration on the melting temperature of ice-like formations [170]. Reproduced by permission of Nauka.

tration of low-molecular-weight compounds compared with that of the polymers (Figs. 3.9 and 3.10).

This distinction seems to be caused by the type of formed bonds. In the case of the analogues, carbonyl and amine groups may both be involved in interactions with H_2O. The amide molecules are connected to water molecules by hydrogen bonds through carbonyl groups. On the other hand, these molecules possess a significant dipole moment and can enter into dipole–dipole interactions with water molecules.

In the case of the polymer, only the formation of hydrogen bonds $C{=}O \bullet\bullet\bullet$ $H{—}O{—}H$ is possible since positive-charged atoms of the amide group are in the vicinity of the main chain and are inaccessible to water molecules due to steric hindrances (IR spectroscopy [120]).

Thus, the macromolecules of poly-*N*-vinylamides, to a variable degree depending on the chemical structure, affects the structural organization of water associates at freezing (or melting), creating ice-like formations with a crystal lattice, distinguished from a crystal lattice of ice in bulk water.

In order to understand these factors, lowering the enthalpy of melting of the ice-like structures in the presence of poly-*N*-vinylamides (PVP, PVCL, PVMA and others), whose chainlinks contain polar amide groups in surroundings of apolar fragments of ethylene main chain and side substituents, let us consider known representations on the structure of ice and water [171, 172] and also data on the heats of melting of ice, made in the presence of water-soluble polymers of another type, namely polymers containing a high local concentration of hydroxyl groups (polyvinyl alcohol) and carboxylic and hydroxyl groups (mucopolysaccharides) [173].

In the hexagonal structure of ice, each water molecule forms two hydrogen bonds acting as a donor of protons, and two hydrogen bonds as an acceptor of protons, thus co-operating with four water molecules, located at the top of a tetrahedron and spaced from the first molecule by a distance of 2.77 Å. The tetrahedral coordination results in the formation of the open-work structure of low density with cavities. Therefore, within the ice structure the H_2O molecules with four hydrogen bonds prevail, and only an insignificant quantity of them forms two or three hydrogen bonds [172].

The heat of ice melting is basically spent on breaking the hydrogen bonds between water molecules that results in reduction of an average number of these bonds per H_2O. However, there is the set of force-non-equivalent hydrogen bonds in crystal. As was established by X-ray analysis [174], each oxygen atom in the framework has four neighbours: one is spaced by a distance of 2.76 Å and three are spaced by 2.94 Å.

Therefore, at heating of ice the insufficiently strong hydrogen bonds (so-called centre-symmetrical bonds (2.94 Å)), forming hexagonal rings, are the first to be destroyed, followed by mirror-symmetrical bonds (2.76 Å) located between layers of hexagonal rings [175, 176]. As a result, a fraction of water molecules remain in ice-like associates with a loose structure (named 'flickering clusters' [177]), reproducing the structure of ice, and the other fraction forms a wide set of water associates of various compositions with a more dense structure consisting of water molecules with a smaller number (less than four) of hydrogen bonds per molecule of water (one, two or three). A wide variety of such associates in bulk water is only the main factor complicating our understanding of all physicochemical peculiarities of this unique liquid.

A large number of bulk water structure models was proposed, and generalized in [171–180]. However, none of these proposed models can in all cases explain the separate phenomena, found in liquid water and attributed to probable structural reorganizations. The picture of interactions in aqueous solutions with molecules of other substances present becomes much complicated. On the other hand, it is necessary to recognize that in a wide variety of intermolecular forces, arising in an aqueous solution, nevertheless the fundamental role is played by the structure of water associates.

Some special features of water associates were established by an advanced half-empirical quantum-chemical method PM3 (modified AMPAC software with the expansion of number of orbitals and the reduction of calculation time).

Table 3.2   The heats of water associate formation and its dipole moments

Water structure models	Number of water molecules	$-\Delta H$ (kJ / mol)	Dipole moment (D)
Cluster	5	17.9	1.0
Chain	3	15.5	2.71
Chain	5	16.2	5.52
Chain	10	17.4	5.81

Two types of water structure were used. The first type is a cluster of five water molecules, one central with four hydrogen bonds and four peripherals bound to the central molecule through one hydrogen bond. The second is an associate being a chain in which each subsequent molecule interacts with the previous one through a hydrogen bond, i.e. each molecule forms two hydrogen bonds (Table 3.2).

It follows from the data that the designed dipole moment of a cluster is rather low ($D = 1.0$), though D of an isolated water molecule is 1.8. In fact, as is well known [178], the permittivity of ice ($\varepsilon = 3.1$) where each water molecule forms four hydrogen bonds in a hexagonal ring, is much reduced in comparison to that of water, which is 87.9 at ~ 0 °C, 78.2 at 25 °C and 55.7 at 100 °C.

Such a sharp difference of permittivity (ε) in going from solid state to liquid ($\varepsilon = 3.1$ to 87.9) seems to be caused by the partial destruction of weaker centre-symmetrical (inside rings) hydrogen bonds, while the mirror-symmetrical ones remain, being much stronger, and create short chains. This chain-like structure of 'twisted threads' due to the structure of water molecule (105° 3' angle of H—O—H) can take place in cavities of clusters (remaining hexagonal or partially deformed hexagonal rings). Thus such structural transformation in the process of phase transition is likely to favour the increase of system density and coordination number per water molecule [171–178].

Quantum-chemical calculations show different heats of the formation of two types of water associates and considerable distinction in their dipole moments (vector sum of dipoles of all bonds) with the increase of molecules number in these associates (Table 3.2).

The structure of an ice-like cluster (one surrounded by four) is favourable for the formation heat, but obtains a low dipole moment ($1.0D$), while at the chain arrangement of water molecules (two hydrogen bonds) both the formation heat and dipole moment grow with the increase of the number of water molecules. It is particularly remarkable that the value of the dipole moment of the associate of the second type grows from $2.71D$ (at three water molecules) to $5.52D$ (at five molecules) and then changes insignificantly up to $5.8D$ at further water molecule accumulation from 6 up to 10.

It is well known [171, 172] that the permittivity (ε) of a liquid is defined not only by the dipole moment of molecules comprising the former but also by the dipole moment of molecular associates. Within these associates such a spatial arrangement of molecules may be achieved that will result in an increase of dipole moments.

Analysis of theoretical calculations of various models of water structural organization (cluster and chain-like with two hydrogen bonds per molecule) (Table 3.2) permits the assumption that structural transformations will take place at the transition of ice into liquid water. In a liquid of high ε which is water at ~ 0 °C the associates with the highest dipole moment should be created. One can assume that these associates are to be made probably of five or six molecules of water (Table 3.2).

In this connection attention is focused on protic liquids of other structures, such as N-monosubstituted amides being in *trans*-form (see Chapter 1). In such liquids the chain associates of high dipole moments appear with hydrogen bond formation, providing high permittivity.

Theoretical calculations of dipole moments of the two models of water structure allow the assumption that the sharp growth of ε from 3.1 (solid state ice) up to 88 at melting (liquid water at ~ 0 °C) is caused by the destruction of hydrogen bonds in a hexagonal ring, creating conditions for the formation of 'flickering', twisted, chain-like, short-distance associates with high dipole moments, consisting of four or five molecules and filling the hollow space in the remaining clusters of hexagonal rings.

Therefore the action of poly-N-vinylamide macromolecules with increasing polymer concentration in water on melting heat of ice-like crystal formations, arising at water freezing in a polymer network, promotes the formation of the chains in which incomplete realization of water molecule hydrogen bonds occurs. At high polymer concentration (> 2.0 mol/l), as Figs. 3.8 and 3.9 suggest, there is no 'free' water to be converted into ice of common structure at freezing.

The point is that the apolar polyethylene framework, apolar side substituents at amide groups and polar carbonyl groups of a large partially negative charge on an oxygen atom accessible to the formation of hydrogen bonds with protons of water molecules, limit the construction space of voluminous hexagonal ice structure and initiates a new arrangement of water molecules near polymer chains.

The mutual approach of apolar sites that virtually do not co-operate with water molecules at the increase of polymer concentration and their polarization by a large $C{=}O$ dipole, the action of which extends not only to the first molecule connected to it but also to others co-operating with the first, and transforms the network of hydrogen bonds in aqueous solution. Then the fraction of the 'threadlike' chain associates will grow, in which not only water molecules with various numbers of hydrogen bonds integrate, but also molecules of other charge distributions induced by carbonyl dipoles.

Special attention must be given to the fact that in the presence of polymers investigated with amide group the significant lowering of melting heat of a frozen aqueous solution with polymer takes place at 16–20 wt% of polymer (230–240 J/g against 334 J/g for pure ice). At the same time, the melting heat of ice with polymers, containing hydroxyl groups (polyvinyl alcohol with 20.7 mol% sulfonic acid groups) and both hydroxyl and carboxylic groups (mucopolysaccharides, as chondroitin sulfate, chondroitin, hyaluronic acid and heparin) [173] differs slightly (300–360 J/g) from that of bulk water, though the polymer concentration (30–40 wt%) is higher than in the case of PVP, PVCL or PVMA. The data do not demonstrate any significant deterioration of the structure of water associates being surrounded by macromolecules of hydroxyl-containing polymers without a decrease of the number of hydrogen bonds per water molecule. It is likely to be stipulated by the polymer composition, where hydroxyl and carboxylic groups act simultaneously as donors and acceptors of protons for water molecules.

As follows from the concept stated above, under the influence of dipoles of poly-*N*-vinylamides and their apolar fragments, the structure of bulk water breaks and redistributes into linear 'threadlike' associates, which begin with the atom of oxygen of C$=$O in the field of low permittivity and expand on a number of water molecules in the solution direction and on a set of types of molecules with various numbers of hydrogen bonds.

Quantum-chemical calculations of hydrogen bond formation energy of water molecules in a chain beginning with C$=$O of *N*-isopropylpyrrolidone (PVP analogue), at consecutive connection (in two ways, see water chain structures 1, 2) of these molecules to one another forming two hydrogen bonds per molecule, show that the first method of connection is more energetically favourable compared to the second one. It is also important that the formation energy of the first hydrogen bond between C$=$O and H$_2$O (12.8 kJ/mol) is smaller than that of the second (22.9 kJ/mol) and the rest bonds between water molecules. In addition the atoms of water molecules have a different charge distribution, being included in this chain (structures 1 and 2)

The lower value of enthalpy of interaction between C$=$O$\cdots$H$_2$O in comparison with that of H—O—H$\cdots$H—O—H is confirmed in refs [179, 181], indicating that the interaction between the anion and H—O—H is weaker than that between water molecules. Moreover, it is necessary to note IR spectroscopic data [120] on the increase of stretching vibration frequencies (v_1 and v_3) of

N-isopropylpyrrolidone (chain structure 1)

H$_3$C—C—CH$_3$ with H above C

N 0.28 —0.39 —0.424 —0.425 —0.443 —0.424 —0.410

0.237   0.234   0.234   0.238   0.231

$=$O$\cdots$H—O$\cdots$H—O$\cdots$H—O$\cdots$H—O$\cdots$H—O

0.193   0.194   0.197   0.191   0.18

H   H   H   H   H

ΔH kJ/mol = 12.8   22.9   20.1   24.3   16.02

N-isopropylcaprolactam (chain structure 1)

$$\text{H}_3\text{C}-\overset{\overset{\displaystyle\text{H}}{|}}{\underset{|}{\text{C}}}-\text{CH}_3$$

	0.238		0.233		0.234		0.235		0.216	
−0.42		−0.423		−0.432		−0.44		−0.423		−0.410
N≡O	---H—O	----H—O	---H—O	---H—O	--H—O
	0.19	0.194	0.197	0.187	0.172
	H	H	H	H	H

$\Delta H_{kJ/mol} = 12.2 \qquad 22.4 \qquad 21.4 \qquad 23.2 \qquad 17.5$

N-isopropylpyrrolidone (water chain structure 2)

$$\text{H}_3\text{C}-\overset{\overset{\displaystyle\text{H}}{|}}{\underset{|}{\text{C}}}-\text{CH}_3$$

N≡O ---H—O—H ----O—H ---O—H ---O—H ---O—H

$\Delta H_{kJ/mol} = 11.7 \qquad 16.35 \qquad 16.06 \qquad 14.3$

D_2O molecules, interacting with C=O of PVP (hydration degree is 0.2–2), in comparison with those of pure water. These changes of v_1 and v_3 of D_2O present in PVP film give convincing proof that the power of the hydrogen bond between polymer C=O and D—O—D is significantly reduced in comparison to that of D—O—D⋯D—O—D interaction, becoming even weaker with the increase of a local concentration of amide groups relative to water content.

Theoretical calculations of charge distribution on atoms of water molecules polarized by dipole and connected in the chain serve as an addition to the previous calculations of other researchers, showing that at dimerization of water molecules a change of charges occurs [171, 172, 177], changing in turn the acidity of hydrogen and basicity of oxygen [172, 177].

Thus, an appreciable reduction of melting enthalpy of ice-like crystal formations generated at freezing of water in the presence of PVP, PVCL and some other polymers is caused by a structural reorganization in water associates under the influence of dipoles near to apolar fragments of macromolecule. In an aqueous polymer solution at rather high concentration of polymer (> 2 mol/l), apparently the conditions are unfavourable for the occurrence of hexagonal ice structures. On the basis of data received by the DSC method and theoretical calculations of some structural models of water associates it is possible to state that the occurrence of water threadlike associates with a lower degree of hydrogen bond realization per water molecule (1, 2 or 3) takes place in the polymer solutions.

In fact, information confirming this conclusion was received on researching water in poorly swollen gel by the DSC method, where it is possible to organize an even stronger influence of specified fragments of polymer chain on water structures.

PHASE TRANSITIONS OF WATER IN PVCL GELS

The PVCL gel is a unique object for the observation of structural transformations within water associates under the action of dipoles in an apolar environment following change of water content in gel [181, 182]. The choice of PVCL for production of hydrogel was also determined by the unusual character of the polymer effect on melting thermograms of ice with macromolecules (Fig. 3.7) [170], where only one unimodal maximum due to the endothermic effect of melting of ice-like formations within a PVCL network is observed at a rather low concentration of PVCL (0.15 mol/l), unlike the thermograms of ice with PVP, PVMA or PVMF.

Solvation (or hydration) within the gel obtained by VCL radical polymerization in the presence of a crosslinking agent (divinylbenzene) can be easily regulated by varying solvent contents and degree of crosslinking. As a result water molecules are subjected on all sides to electrostatic fields created by C=O dipoles within a slightly polar medium.

Analysis of the data concerning phase transitions (ice melting and evaporation) permits stronger transformations of water associate structure at the penetration of water into PVCL gel channels. This material has the following chemical structure:

$$
\left[\begin{array}{c} -CH_2-CH \\ \\ CH_2\diagup N\diagdown C=O \\ / \hspace{1.2cm} \backslash \\ CH_2 \hspace{0.6cm} CH_2 \\ \backslash \hspace{0.9cm} / \\ CH_2-CH_2 \end{array}\right] \left[\begin{array}{c} CH_2-CH \\ \\ \text{(benzene ring)} \\ \\ CH_2-CH- \end{array}\right]_\beta
$$

$(\beta = 1, 5; 3, 5; 5, 0 \text{ mol\%})$

The polymer effect on the melting character of ice-like formations within PVCL–DVB gel ($\beta_{DVB} = 3.5$ mol%) is clearly defined on thermograms of pure ice and frozen water-gel (Fig 3.11). The effect of gel swelling degree (or the number of H_2O per chainlink (N)) on the thermogram curve form is demonstrated on Fig 3.11. It is seen that in all cases two endothermic peaks fall in a range of temperatures from $-10\ °C$ to $+15\ °C$ depending on N. At $N = 15$ a bimodality of thermal effect appears, which is exhibited by two wide maxima in low-temperature (t_1) and in high-temperature (t_2) ranges.

Concerning the temperature range where the take-up of melting heat is observed, one could relate the low-intensity, low-temperature peak to a bulk water ice melting. However, the presence of only one peak on the thermograms of frozen aqueous PVCL solution melting (Fig 3.7) and a low degree of gel swelling (60–190%) permits the absence of free (bulk) water within the gel.

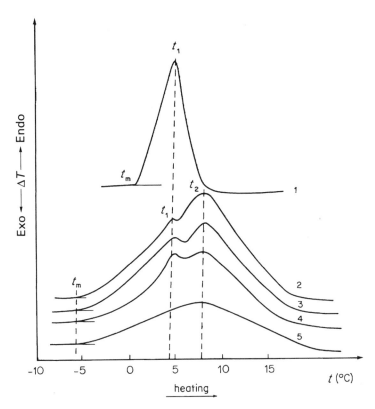

Fig. 3.11. The heating thermograms of bulk water ice (1) and ice-like formations within PVCL–DVB gel (β_{DVB} = 3.5 mol%) at various number of water molecules per chainlink (N): 14 (2); 12 (3); 8.6 (4) and 4.9 (5). Heating rate is 5 °C/min [181, 182]. Reproduced by permission of Nauka.

This conclusion is also confirmed by melting thermograms of frozen water within PVCL–DVB of β_{DVB} = 5.0 mol%, where the two peaks change their height with the decrease of N (from 12 to 3.3) (Fig. 3.12).

At melting of ice-like formations within PVCL–DVB of β_{DVB} = 5.0 mol% at low water contents (N = 3.3), as follows from Fig. 3.12, the high-temperature peak (t_2) disappears and only the low-temperature one remains. At further heating a new endothermic peak (t_5) is displayed in a temperature range of 35–65 °C, with a maximum at 47 °C. It is necessary to note the absence of this peak for all other gels (PVCL–DVB with 1.5 and 3.5 mol%) investigated and at various degrees of hydration. A similar peak arises for the sample of PVCL–DVB of β_{DVB} = 5.0 mol% at N = 6. A further increase of water molecule number per chainlink ($N > 8.5$) forces this peak to disappear.

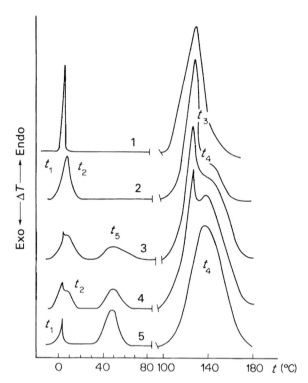

Fig. 3.12. The effect of water contents (*N*) on the form of heating thermograms of frozen water–gel systems with PVCL–DVB of β_{DVB} = 5.0 mol%. Curve 1, pure water; Curve 2, *N* = 11; Curve 3, *N* = 8.5; Curve 4, *N* = 5.5; Curve 5, *N* = 3.7. [182]. Scan rate is 5 °C/min.

The phase transition found is accompanied with squeezing out of some water as thermogravimetric measurements attest to the plausible reduction of system weight at heating from 35 °C to 70 °C. It is important to emphasize the absence of any thermal effects due to, for example, conformational transitions in matrix, in a dry sample of PVCL–DVB (β_{DVB} = 5.0 mol%), in such a temperature range where those effects could be felt. In other words, there are fields of various (at least, three types) surroundings of water associates in a gel network at certain contents of water molecules (*N* < 10) at *T* > 0 °C.

Moreover, the two endothermic peaks are observed at water evaporation out of the gel at *t* > 95 °C. Both the first, low-temperature peak (t_3) and the second, high-temperature peak (t_4) represent the expenditure of energy due to water evaporation. It is remarkable that their intensities depend on the initial water contents (*N*) and are redistributed with the decrease of *N*. The low-temperature peak is highly intensive at *N* = 11–12, becomes reduced at *N* = 5–7 and almost disappears at *N* < 5, whereas the other one grows significantly.

Fraction of high-temperature peak area (%)

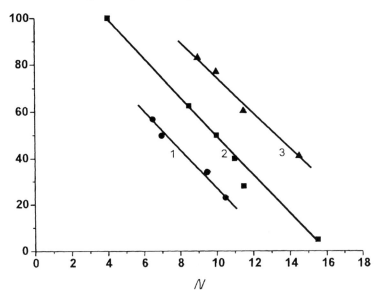

Fig. 3.13. The dependencies of ratio of the high-temperature peak area to the total area of the thermograms of water evaporation out of gels made from PVCL–DVB, $[\beta_{DVB} = 1.5(1);$ $3.5(2); 5 \text{ mol}\%(3)]$ on the water molecule number per chainlink (N) [182].

It follows from Fig. 3.12 that the cymbate growth of three peaks (t_1, t_4, and t_5) takes place on thermograms in the range of melting temperatures of ice-like formations and in the range of evaporation temperatures of water associates. At the same time the increase of intensity of the specified peaks is accompanied by a fall in intensity of the high-temperature peak (t_2) at melting and the low-temperature peak (t_3) at evaporation. The intensity growth of the high-tempera-ture peak in the range of evaporation temperatures (t_4) at the decrease of N is followed from the ratio of the peak area to the total area of water evaporation thermogram of all three samples of various crosslinking degree (Fig. 3.12).

As the enthalpy of hydrogen bond formation between water molecules and $C{=}O$ is reduced in comparison with that of the bond between $O{-}H \cdots O{-}H$ in accordance with IR-spectroscopic data of PVP$\cdots$H$-$O$-$H [120] and quan-tum-chemical calculations, the low-temperature peak on the water evaporation thermogram (t_3) can be attributed to the evaporation of water molecules poorly bound to $C{=}O$ in gel. Water molecules near to apolar groups, promoting water associate structuring, evaporate at higher temperatures and are responsible for the occurrence of the high-temperature peak (t_4).

These peculiar characteristics of endothermic processes of phase transitions of water within gels of PVCL–DVB, as deduced from DSC measurements, are due to the specific structural organization of water associates in a slightly swollen polyamide matrix. So, the occurrence of two peaks on heating thermograms of frozen water in the hydrogel PVCL–DVB (β_{DVB} = 3.5; 5.0 mol%) is stipulated by the phase transition of the first kind (melting) of ice-like formations of water molecules (1), the crystal structure of which was generated near apolar fields of methylene groups of ethylene chain and caprolactam rings being in 'chair' conformations (narrow low-temperature peak t_1), and ice-like formations made of polarized water molecules under the action of C=O dipoles (2) (high-temperature peak t_2).

The appearance of the new peak (t_5) at $N < 8.5$ in a more dense gel during the heating process indicates the presence of some water molecules in water associates, located in a space between caprolactam rings and having an unlike structure probably like that of clathrate hydrates of low molecular organic compounds. The portion of such a type of water in comparison with the total water amount in gel is small (10–20%) and dependent on the initial number of molecules (N). Therefore, at N = 3.7, 5.5, 8.5 and 11 this fraction makes 20, 16, 11 and 0% respectively. Probably at such water contents inside the gel there are some separate locations of a specific arrangement of rings and that of water molecules near to the ones. At the further removal of water molecules a conformational mutual approach of some rings can occur, resulting in structural disposition similar to VCL crystal structure (see Chapter 1). It will be recalled that in a crystal cell consisting of two rings (in 'chair' conformation), the rings are arranged so (as two chairs combined by seats) that the greatest dispersion and dipole interactions between them are realized.

Accordingly, in the other type of water surrounding, the dipoles of polar amide groups play a crucial role. The evaporation of water molecules from these associates proceeds actively because of the opportunity for H_2O to move on C=O groups as 'stepping stones' (narrow low-temperature peak at t_4).

In the spaces between apolar fragments (ethylene chain and methylene groups of rings) water associates arise where water molecules are structured and bound to themselves. Such surrounding and structural organization of water associates seem to be responsible for the occurrence of a wide high-temperature peak (t_4).

Further attention should be paid to the fact that the value of the initial temperature at which the endothermic effect is displayed (the melting temperature of ice-like formations, T_{mlt}) in PVCL gel at various N and at β = 1.5 and 3.5% remains almost constant, being –8 °C. In the case of PVCL–DVB (5 mol%) the value of T_{mlt} is 5–6 °C, whereas in the aqueous PVCL solution (C_{POL} = 2 mol/l) it is reduced to a greater degree, being – 12 °C (Fig. 3.8).

At the earlier studies of phase transition (melting) of water in hydrogels based on copolymers of VP-methylmethacrylate (MMA) [183–185] it was suggested that water exists in several states: 'free' water forming the structure of the ice in bulk water, and water 'bound' to polymer, crystallizing in ice-like formations. The presence of 'non-freezing' water, not forming an ice-like crystal lattice, is also assumed.

Comparison of T_{mlt} values of frozen (so-called 'bound') water in polymers of various structure, e.g. in crosslinked PVCL–DVB [181, 182], in PVCL solution [170], in mucopolysaccharides (chondroitin, chondroitinsulfuric acid, heparin, etc. [173]) and in hydrogels based on VP–MMA copolymers [183–185] allows the essential conclusions to be made.

Thus, ice-like formations of frozen PVCL solution (26 wt% of PVCL) melt at -12 °C, though in the case of slightly swollen crosslinked PVCL with much greater polymer chain contents (50–65 wt%) melting starts at -8 °C and above.

In the case of mucopolysaccharides at approximately the same number of H_2O molecules per chainlink as in crosslinked PVCL ($N = 10$–15) and ice-melting heat in the saccharide chainlinks surrounding water molecules close to that of ice in bulk water, T_{mlt} decreases even further (-10° to 15 °C) [173].

For hydrogels of VP–MMA copolymers of various composition at the equilibrium water contents the value of T_{mlt} of 'bound' water depends in a complex manner on VP contents in the copolymer, being -10 °C at 37.8 mol% of VP, -7 °C at 52.6, -4 °C at 66.8 and -8 °C at 83.1 mol%.

Analysis of these data relating to T_{mlt} of ice in diverse polymeric systems manifests that the range of T_{mlt} fall for ice-like formations generated in the presence of water-soluble polymers, in comparison with that of pure ice can be characterized not only by the sizes of ice-like formations [186] but also by the nature of functional groups surrounding these associates and also by the spatial arrangement of fragments near to which H_2O molecules are placed.

It should be noted that at $T_{mlt} = -8$ °C exceeding both T_{mlt} of ice-like formations with non-crosslinked PVCL macromolecules (-12 °C) and that of ice within mucopolysaccharide networks (-10 °C, -17 °C) there is a significant decrease of ice melting heat (ΔH_{mlt}) depending on the initial contents of water (see Fig. 3.14). With the decrease of N from 15 to 5, ΔH_{mlt} decreases from 180 to 54 J/g. It is essential that these ΔH_{mlt} values are much reduced in comparison with those for ice with non-crosslinked PVCL (26 wt%) (230 J/g) and of ice with mucopolysaccharides (320–330 J/g [173]).

In this case the low value of ΔH_{mlt} of ice-like formations within the PVCL network can probably be attributed to the fact that at freezing of water associates under the action of amide groups and apolar fragments the ice-like formations are organized, their lattice being constructed with the participation of H_2O molecules with a smaller number of hydrogen bonds. Melting of such ice-like crystals requires less expenditure of energy due to the smaller number of hydrogen bonds, less than four per H_2O, than in the case of pure ice with crystal lattice.

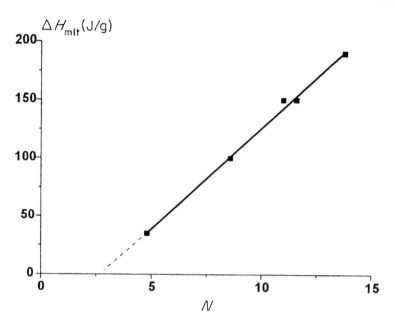

Fig. 3.14. Dependence of total melting enthalpy of ice-like formations in crosslinked PVCL ($\beta = 3.5$ mol%) on water contents in hydrogel (N) [182].

As seen on Fig. 3.14, an extrapolation of the straight line until its intersection with the abscissa axis gives an average number of water molecules (about two) incapable of forming a crystal lattice, i.e. 'non-freezing' water.

Earlier it was observed that the average number of H_2O molecules per chain-link in copolymer attributed to 'non-freezing' water, in VP–MMA hydrogels [180] is five to seven and exceeds that of PVCL gels. One can believe that the presence of 'non-freezing' water incapable of ice-like structure organization in VP–MMA copolymers is caused by association of these molecules not only close to C=O of VP rings but also close to C=O of MMA. Carbonyls of MMA being dipoles of a low dipole moment ($2.5–2.7D$) can also polarize adjacent water molecules.

The value of ΔH_{mlt} (Fig. 3.14) is minimal at just two to three water molecules per chainlink. As a result, water associates of two molecules being in different locations in the chain do not 'overlap' each other and cannot form ice-like crystals.

In a swollen hydrogel ($N > 5–6$) the action C=O of the amide group on these two to three water molecules and through them probably organizes threadlike extended chains involving other water molecules due to steric restrictions, created by voluminous rings and high local concentration of amide groups. The

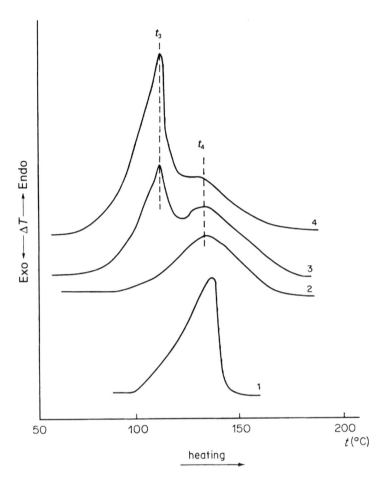

Fig. 3.15. The evaporation thermogram of pure water (1), water from a PVCL–DVB gel of $\beta_{DVB} = 3.5$ mol% at various N: 4.9 (2), 8.6 (3) and 11.7 (4) [181, 182].

influence of dipoles on the association of H_2O molecules, as seen on Fig. 3.14, is retained at quite large distances from a polymer chain, inducing a new structural organization in the surrounding water. The lower ΔH_{mlt} value is retained in ice-like formations inside the PVCL gel at $N = 10-15$ (Fig. 3.12) and at $N = 20-30$ (Fig. 3.9) for non-crosslinked PVCL.

As for the high-temperature phase transition of water (evaporation), the character of the action of water contents in the PVCL–DVB gel of $\beta_{DVB} = 5$ mol% on endothermic effect is shown on Fig. 3.15. A bimodality of the thermal effect is found, expressed by two maxima (low-temperature peak t_3 and high-temperature peak t_4). The occurrence of these peaks, as mentioned above, is defined by two types of surrounding of water associates: polar surrounding (1) created by carbonyl groups and apolar surrounding (2) of chain fragments. This effect is followed in samples of other structure (PVCL–DVB of $\beta_{DVB} = 3.5$ and 1.5 mol%): the heating thermograms show bimodality of endothermic effect (Fig. 3.15).

At the same time the values of H_{evp} calculated from the total area of thermograms at the known water contents in gels under investigation correspond to that of pure water evaporation (2260 J/g), to within experimental error (2100–2200 J/g).

Let us consider once again the peculiarities of water evaporation out of PVCL networks, paying attention to the dependencies of evaporation heat corresponding to the low-temperature peak ($\Delta H_{evp\,lt}$) and the high-temperature one ($\Delta H_{evp\,ht}$) on the initial water contents (N) (Figs. 3.12 and 3.15).

At high N ($N = 15$) water molecules are positioned throughout the whole gel volume and are under the action of amide groups, as in the PVCL solution. Probably the space between apolar groups promotes the organization of thread-like chains of water molecules under the polarizing action of C=O dipoles. These associates include molecules with a low number of hydrogen bonds. As a result of weakening of these interactions, among them the weaker hydrogen bonds between water molecules and carbonyls, water evaporation proceeds actively, causing a low-temperature peak on an evaporation thermogram.

At the decrease of water molecule number in gel ($N < 10$) the role of apolar fragments of caprolactam rings is strengthened, probably due to their specific conformational structure ('chair' conformation). They begin to push away H_2O molecules from C=O groups and structure them in such a manner that the chain-like water associates create in the apolar fields a network of hydrogen bonds among themselves without influence of the dipole. Evaporation becomes difficult due to reinforcing of hydrogen bonds and sterical barriers of a conformational nature.

The action of PVCL crosslinking degree on the process of water phase transitions is exhibited in different contributions of two (or three) types of the local environment of water associates generated in the gels to the peculiarities of ice-like formation melting and water evaporation.

In conclusion it is necessary to emphasize that poly-N-vinylamide macromolecules, in particular PVCL, represent themselves as strong modifiers of water structure both in solution and in hydrogel, causing reorganization in the arrangement of molecules as a result of the polarizing action of C=O dipoles only on protons of water molecules, with the occurrence of molecules with various degrees of hydrogen bond representation in the surroundings of apolar fragments. In turn, water near to polymer chains defines some specific physicochemical properties of macromolecules which are shown only in aqueous solutions.

3.4 CONFORMATIONAL TRANSFORMATIONS OF THE POLY-*N*-VINYLCAPROLACTAM AND POLY-*N*-VINYLPYRROLIDONE MACROMOLECULES IN AQUEOUS SOLUTIONS

POLY-N-VINYLCAPROLACTAM: PHASE SEPARATION AND VISCOSIMETRY

In a number of poly-*N*-vinylamides investigated, PVCL occupies a special place as it possesses the lowest critical solution temperature (LCST) in aqueous solutions in the range of temperatures at which living organisms are functioning [87, 106, 166, 167]. This temperature (the temperature of phase separation $T_{ph.s.}$) of PVCL of $M_w > 50 \times 10^3$ is 32–37 °C (Fig. 3.1).

Let us consider the unusual practice of the demixing of PVCL macromolecules on the introduction of organic solvents of PVCL (themselves being fully soluble in water) in its aqueous solution. It is remarkable that the addition of alcohols or other water-soluble compounds in an aqueous PVCL solution can result in a decrease of $T_{ph.s.}$ (< 30°) [166, 167]. It testifies that the mechanism of PVCL thermoprecipitation in water or in water–organic mixtures is rather complex. As follows from a previous section (3.2), thermoprecipitation of PVCL macromolecules in water can be caused by the action of a large number of factors including the following:

1. Hydrophobic interactions between rings, connected with change of H_2O structure near to methylene groups, and between chainlinks and apolar fragments of entered substances at heating;
2. Hydrogen bonds between carbonyl groups and water molecules;
3. Apolar dispersion interactions between rings in 'chair' conformation;
4. Hydrogen bonds between H_2O molecules and polar groups of the additives;
5. Structural transformation of water associates in a PVCL hydrate shell.

The value of $T_{ph.s.}$ reflects in an indirect manner the balance of these forces acting in PVCL coils. The increase of the temperature of PVCL aqueous solution and the change of solvent nature strengthen or weaken the action of the above factors, influencing the conformational state of macromolecules in solutions.

It was found by a viscosimetric method [87, 106, 166, 167] that the increase of solution temperature from 10 °C to 30 °C is accompanied by significant (approximately twofold) reduction of intrinsic viscosity of PVCL in water. Other poly-*N*-vinylamides (PVP and PVMA) do not display such a phenomenon. The slight reduction of [η] of PVP takes place only at a significant increase of temperature [106].

The appreciable decrease of [η] of PVCL testifies to the shrinkage of polymeric coils accompanied by a partial dehydration. The water at a temperature higher than 32–35 °C becomes a poor solvent of PVCL, but is an effective solvent for PVP, PVMA and PVMF. In fact, the value of α in the empirical Kuhn–Mark–Houwink equation

$$[\eta] = kM_w^\alpha$$

relating intrinsic viscosity and molecular weight PVCL, in water at 25 °C is 0.57 [156]. The water at 32–34 °C becomes θ-solvent ($\alpha = 0.5$) for PVCL macromolecules.

THE INTRAMOLECULAR MOBILITY OF PVP AND PVCL MACROMOLECULES (METHODS OF PARAMAGNETIC LABELS AND POLARIZED LUMINESCENCE)

The intramolecular mobility of macromolecules in solutions was studied by the EPR method [187, 188]. This method assumes isotropic rotations of the label, together with a chain fragment with which it is covalently connected and reflects label dynamics and small-scale mobility of polymer chain fragments [187].

Spin-labelled PVP and PVCL (PVP-L-1, PVP-L-2, PVCL-L-1 and PVC-L-2) were synthesized by a radical polymerization of the appropriate monomer with vinylpiperidine or allylpiperidine with the following oxidation of amine covalently connected to the chain ($\beta_{label} = 1$–2 mol%) to a nitroxyl radical [187, 188].

Characteristic times (τ_{seg}) of segment rotation in PVP and PVCL in ethanol and the *S*-parameter, describing spatial (steric) restrictions of label rotation and

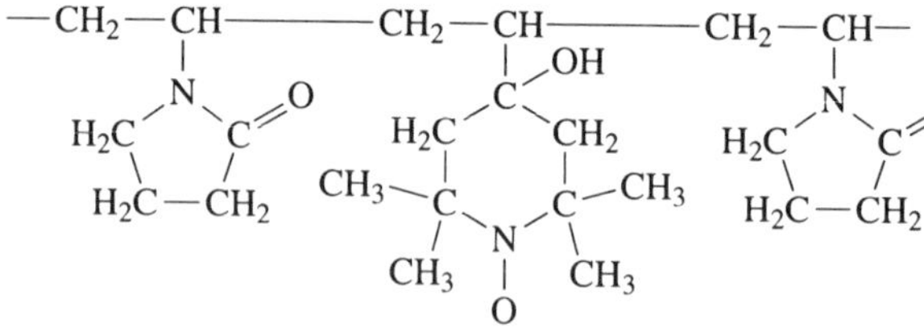

PVP-L-1

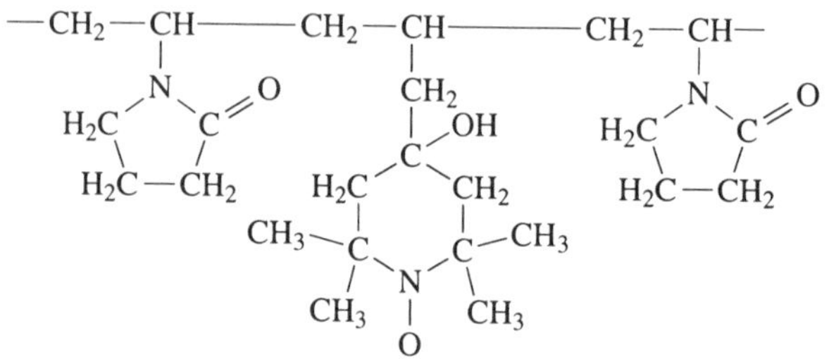

PPVP-L-2

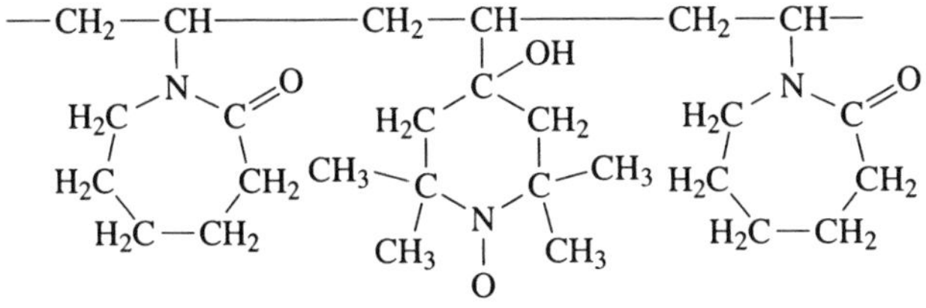

PVCL-L-1

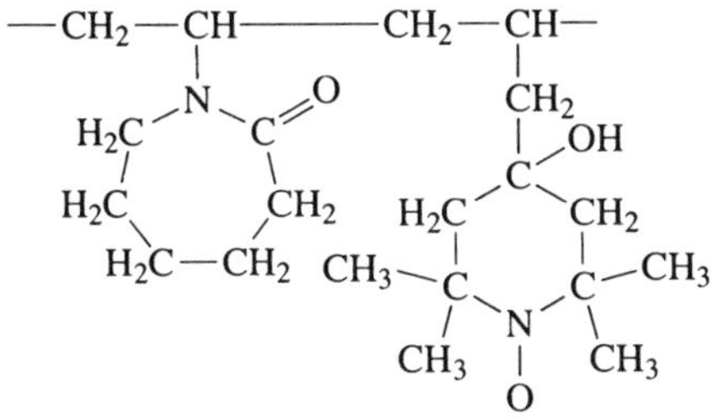

PVCL-L-2

determined by an aperture angle of a 'cone' which the label forms while rotating, are represented in Table 3.3. The S-parameter depends on label structure: the more rigidly the label movement is restricted, the lesser the aperture angle and the closer the value of S to unity. Comparison of characteristic times of segment rotation in PVP and PVCL chains (Table 3.3) shows that the increase of the side substituent size in the chain results in lowering of its segmental mobility, i.e. increase of τ_{seg} [187].

Let us consider the temperature effect on these parameters for spin-labelled PVP and PVCL.

Table 3.3 Parameters of label mobility in PVP and PVCL of various M_w in ethanol (25 °C) [185]. Reprinted from Europ. Polym. J. 19, Wasserman, A. M., Timofeev, V. P., Aleksandrova, T. A., Karaputadze, T. M., Shapiro, A. B. and Kirsh Yu. E., 333–339, © 1983, with kind permission from Elsevier Science Ltd., The Boulevard, Langford Lane, Kidlington OX5 1GB, UK

Polymer	$\bar{M}_w$	$\tau_{seg}m$ (ns)	$S \pm 0.02$
PVP-L-1	73 000	4.4	0.81
PVP-L-2	66 000	4.4	0.74
	23 000	4.4	0.74
PVCL-L-1	41 000	8.8	0.81
PVCL-L-2	52 000	8.8	0.76

Table 3.4 Temperature effect on the intramolecular mobility parameters of PVP-L-1 and PVCL-L-1 in water [189]. Reprinted from *Europ. Polym. J.* **19**, Wasserman, A. M., Timofeev, V. P., Aleksandrova, T. A., Karaputadze, T. M., Shapiro, A. B. and Kirsh, Yu, E., © 1983, with kind permission from Elsevier Science Ltd, The Boulevard, Langford Lane, Kidlington, OX5 1GB, UK

Polymer	Temperature, (°C)	τ_{seg} (ns)	τ_{seg}^{a} (ns)	$S \pm 0.02$
PVP-L-1	1	5.1	2.7	0.83
	10	3.5	2.6	0.80
	20	2.4	2.4	0.75
	30	1.8	2.5	0.70
PVCL-L-1	1	13	6.9	0.84
	10	9.6	6.9	0.82
	20	7.1	7.1	0.81
	25	5.5	6.4	0.83
	30	5.3	6.8	0.84
	32	5.7	8.3	0.86
	36	>1000	—	0.80
	60	>1000	—	0.80

[a] τ_{seg}^{*} is τ_{seg}, determined by the Stokes-Einstein solvent viscosity equation at 20 °C (293 K) and water viscosity at the same temperature.

The increase of polymer solution temperature from 1 °C to 30 °C is accompanied by growth of segmental mobility in PVP and PVCL macromolecules (Table 3.4), and the *S*-parameter decreases in both PVP and PVCL. However, in the case of PVCL it decreases to a smaller degree with temperature increase up to 25 °C, and then starts to increase with increasing temperature up to 30 °C.

As the temperature approaches $T_{ph.s.}$ of PVCL in water, τ_{seg} and the *S*-parameter start to increase (Table 3.4). At $T > T_{ph.s.}$ in an EPR spectrum of the spin-labelled PVCL-L-1, extremely wide peaks appear [184]. The segment mobility in the precipitated system of PVCL differs slightly from that of a solid-state polymer ($\tau_{seg} > 1000$ ns). The increase of the *S*-parameter in a temperature range from 20 °C ($S = 0.81$) to 32 °C ($S = 0.86$), where the intrinsic viscosity of PVCL in water decreases, indicates that before complete PVCL precipitation conformational transformations in polymer coil occur because of intramolecular interactions between the rings, imposing certain spatial restrictions on the label rotation [187, 188].

The data received by a polarized luminescence method using an anthracene label on a chain of PVP (PVP-F), PVMA (PVMA-F), and PVCL (PVCL-F) in solvents of various nature, confirm the greatest slowing down of segmental mobility of PVCL chains in aqueous solutions: the highest relaxation times (τ_{w}^{sp}) were given for a PVCL chain in water compared to alcohols, $CHCl_3$ and DMF (Table 3.5) [190].

Table 3.5 Relaxation times (τ_w^{sp}) characterizing the intramolecular mobility of poly-N-vinylamide macromolecules in various solvents [190]. Reproduced by permission of Nauka

| | τ_w^{sp} (ns)[a] | | | | |
Polymer	Water	Methanol	Ethanol	CHCl$_3$	DMF
PVMA-F	9.4	4.6	—	—	2.8
PVP-F	9.4	4.6	—	—	2.8
PVCL-F	21	7.4	8.2	7.4	2.8

[a] $\tau_w^{sp} = (1/P_0^l + 1/3) \cdot 3\tau_f/(1/P-1/P_0^l)$
where P is the meassured value of luminescence polarization of a solution of the spin-labelled polymer (PVP-F, PVCL-F or PVMA-F with 0.1 mol% of anthranylmethylmethacrylate), τ_f the duration of luminescence of a fluorescent label and $1/P_0^l$ the parameter corresponding to the amplitude of the label travel.
  The change of τ_w^{sp} value due to the change of specific solvent viscosity (η_{sp}) is determined by the following equation: $\tau_w^{sp} = \tau_w (\eta_{sp}/\eta)$.

PVP-F

It follows from the data on polymer behaviour in various solvents (Table 3.5) that the intramolecular mobility is slowed down appreciably in aqueous solutions compared to organic media. Possible reasons for the greater slowing down to PVCL chainlink travel are specific chain microstructure (syndiotactic sequence in PVCL chain) and modified hydrate shell, favouring intramolecular interactions between voluminous rings in 'chair' conformation.

The difference of intramolecular chain mobility in PVCL in comparison with other poly-N-vinylamides is clearly shown at changing composition of solvents mixture (water–ethanol, water–DMF) (Fig. 3.16).

As seen in Fig. 3.16, the value of τ_w^{sp} in the case of PVP lowers the already insignificant content of organic solvent in water. On the other hand, in the case of PVCL, mobility stays constant in water–alcohol mixtures with alcohol content up to 14.5 mol%, and then starts to increase sharply when the contents range from 14.5 to 30 mol%. Thus relaxation time falls from 21 to 6 ns.

The plateau in the slope in the case of PVCL, together with its absence in PVP (or PVMA) testifies to some internal structural organization in aqueous PVCL

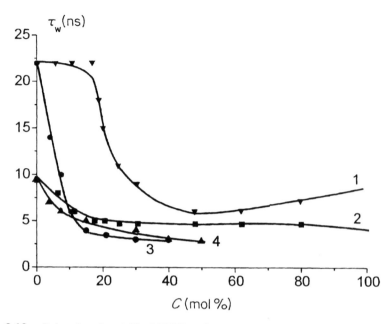

Fig. 3.16. Relaxation time (τ_w^{sp}) of PVCL (1,3) and PVP (2,4) in water–ethanol (1,2) and water–DMF (3,4) mixtures (mol% of organic solvent, η_{sp} = 0.38 mPa s, 25 °C, 0.05 wt%) [190]. Reproduced by permission of Nauka.

coils [190]. Most likely the introduction of alcohol (ethanol or methanol [106]) into water in a specified concentration range (up to 14.5 mol%) is not capable of changing a hydrate surrounding which is generated from water molecules close to the PVCL chain under the action of dipoles and apolar fragments (see section 3.2). The following addition of ethanol to water causes this structure to disintegrate and alcohol molecule enters this shell affecting the interaction of water molecules with carbonyl groups.

This structure of PVCL–hydrate shell deteriorates (Fig. 3.16) in the presence of DMF more rapidly than under the action of alcohol. It means that DMF molecules which are highly capable of strong hydrogen bond formation interact with polarized molecules near the chain, bringing about the destruction of the hydrogen bond network and, as a result, of the above-mentioned structure in PVCL coils. This DMF action raises the intramolecular mobility of PVCL fragments (τ_w^{sp}) in the presence of a small amount of DMF (5–10 mol%) in aqueous PVCL solution.

The peculiarities of hydration of PVCL macromolecules in comparison with that of PVP and PVMA are found in the substantial growth of τ_w^{sp} at heating of

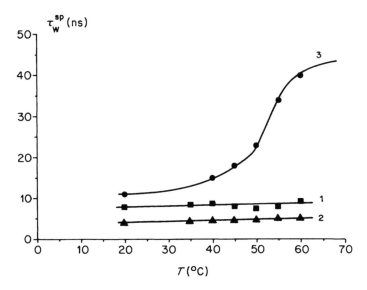

Fig. 3.17. The temperature effect on relaxation times (τ_w^{sp}) of PVP in water (1), PVP in water + DMF (24 wt%) (2) and PVCL in water + DMF (24 wt%) (3) [190]. Reproduced by permission of Nauka.

aqueous solutions of poly-N-vinylamides (Fig. 3.17). Thus τ_w^{sp} of PVCL (even in the presence of 24 wt% DMF) gradually increases from 9 to 40 ns as temperature increases from 20 °C to 50–60 °C [190]. The relaxation times of PVP solution are not significantly affected.

The data concerning the intramolecular mobility of PVCL chain fragments in water on the addition of ethanol correlates with that on hydrodynamic behaviour of coils and change of phase separation point of PVCL in mixtures (Fig. 3.18) [166, 167].

Ethanol, being a fine solvent of PVCL, is incapable of destroying a specific water shell near to PVCL chain until its content reaches 14.5 mol%. Probably at this concentration ethanol molecules fail to enter into immediate contact with $C{=}O$ groups, forming hydrogen bonds due to a specific hydrate shell repelling the alcohol molecules.

The data of Fig. 3.18 concerning the slight decrease of [η] and $T_{ph.s.}$ at an ethanol concentration ranging from 1 to 14.5 mol%, seem to reflect the action of ethanol on hydrated layers being apart from the $C{=}O$ groups of PVCL macromolecules in solution. The introduction of ethanol up to 14.5 mol% has no effect on a hydrate layer situated near to the PVCL chain which is proved by the constancy of τ_w^{sp} in the same concentration range of ethanol (Fig. 3.15). In other

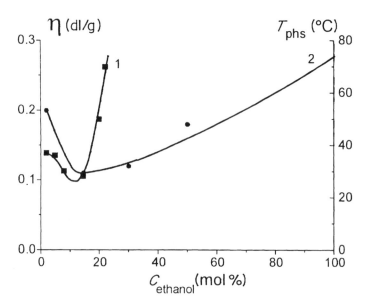

Fig. 3.18. Concentration effect of ethanol in water on intrinsic viscosity (1) and phase separation temperature point (2) of PVCL ($M_w = 27 \times 10^3$) [166]. Reproduced by permission of Nauka.

words, ethanol molecules can partially replace water molecules, being apart from the main chain. The solvate (ethanol+water) layer is connected to the macromolecule through a polarized water molecule layer.

With the further introduction of ethanol ($C_{eth} > 20$ mol%) alcohol molecules start to break through the water shell to C=O groups, creating a new solvate surrounding the nearby PVCL chain. This process is accompanied by a sharp growth of $T_{ph.s.}$ and a marked increase in intrinsic viscosity (Fig. 3.17).

CONFORMATIONAL TRANSFORMATIONS OF PVCL MACROMOLECULES IN WATER (BY NMR DATA)

In the low-temperature range ($< 10\,°C$) PVCL macromolecules are in an expanded conformational state, affecting a large volume of the surrounding water and modifying conditions of ice crystal structure formation (see section 3.2).

The increase of temperature from 0 °C to 32–33 °C ($T_{ph.s.}$), promotes the shrinkage of PVCL coils that is expressed in the decrease of $[\eta]$ [87, 166, 167] (Fig. 3.20), segmental mobility (τ_w^{sp}) [106, 188, 189] and in the growth of the S-parameter of rigidity.

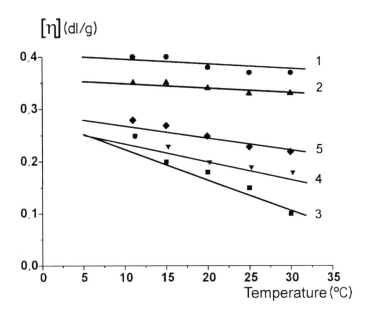

Fig. 3.19. Temperature effect on intrinsic viscosity of PVP (1) PVMA (2), PVCL (3), VP–VCL copolymers of 81.3 mol% (4), and 29.1 mol% of VCL (5), in water [166]. Reproduced by permission of Nauka.

The intrinsic viscosity of PVCL as well as that of VCL–VP copolymers decreases considerably as the temperature rises from 5 °C to 30 °C (Fig. 3.19). This effect is not as strong in PVP and PVMA solutions. PVCL coils in dilute solutions become more compact and tight since a fraction of water molecules is forced out of the hydrate shell by intramolecular (dispersion and dipole–dipole) interactions between voluminous rings, together with hydrophobic interactions caused by the change of water associate structure.

It is important that with the increase of temperature the chemical shift and half-width of the C=O peak in the ^{13}C NMR spectrum of aqueous PVCL which characterize the amide hydration, changes significantly as the temperature approaches $T_{ph.s.}$. The peak is shifted upfield (0.3 ppm) and this effect is greatly enhanced at $T_{ph.s.}$ when polymer–water phase separation occurs. This phenomenon is caused by partial dehydration of C=O affecting the charge value on oxygen and carbon atoms in C=O. Enhancement of the peak's half-width indicates considerable restriction of segmental chain mobility (Fig. 3.20).

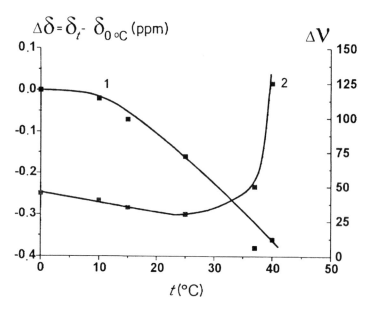

Fig. 3.20. Dependence of chemical shift (1) and half-width (2) of the signal of carbon atom of C=O in ^{13}C NMR spectrum of PVCL on water temperature at 1.6 mol% of PVCL [169]. Reproduced by permission of Nauka.

3.5 TEMPERATURE-SENSITIVE PHASE TRANSFORMATIONS OF PVCL IN WATER–ORGANIC MIXTURES

LOCAL DENSITY OF CHAINLINKS IN A COIL OF FLEXIBLE-CHAIN POLYMERS IN A DISSOLVED CONDITION

A number of physicochemical properties of flexible-chain polymers is defined by concentration distribution of chainlinks in a coil due to the chain structure of macromolecules [191]. In order to thoroughly understand the phenomena appearing in aqueous and water–organic solutions of poly-N-vinylamides, one must pay attention to the local concentration of chainlinks in close proximity to a certain chainlink. The point is that the local concentration can be appreciably greater than the average density of chainlinks in the volume of the whole polymer coil. This factor is responsible for the conformational state of macro-molecules in solution, especially when some additional interactions occur between side substituents in chainlinks, e.g. hydrophobic interactions, hydrogen bonds, dispersion and dipole–dipole interactions or other.

The local density of chainlinks was determined by the method of spin labels using poly-4-vinylpyridine as an example of a flexible-chain polymer in good solvent (ethanol) [188, 189]. As the α-parameter of poly-4-vinylpyridine in the equation $[\eta] = KM^{\alpha}$ is 0.6–0.7, which is close to that of PVP in water and of PVCL in alcohols, one can believe that existing local density of chainlinks in poly-4-vinylpyridine does not differ much from that in PVP chains and other poly-N-vinylamides.

$$\beta = n/(m + n) = 0.025, 0.05, 0.1, 0.2 \text{ mol}\%$$

The determination of local density of chainlinks is based on the analysis of EPR spectra of spin-labelled macromolecules with a large number of labels. Dipole and exchange interactions between labels appear in these EPR spectra. The approach developed in ref. [187] allows the estimation of the local density of chainlinks in a macromolecule coil, considering dipole interactions if the number of labels on a polymer chain is well known, and to calculate the rate constants of intramolecular collisions of labels and coefficients of their local transmitting diffusion, considering the exchange interactions.

The radius of a local spherical area in which the local density of chainlinks is determined is ~30 Å, and meets at a distance on which the dipole contribution to the line width of a central label from labels on the border of the sphere matches the width of a line corresponding to 0.5–1.0 Hz. As a rule the elementary volume in which dipole interactions of labels are displayed is less than the volume of a polymer coil.

The local density of chainlinks in a poly-4-vinylpyridine chain was estimated using the dependence of the width of EPR lines on spin-label constants of macromolecules (Fig. 3.6).

It follows from the data in Table 3.6 that for the chains of poly-4-vinyl-pyridine with large M_w the local density exceeds four to six times the average density in coil volume, and virtually does not depend on polymer concentration in solution (up to 2 mol/l) [192, 193]. Therefore, one can assume that in close proximity to any chainlink within poly-N-vinylamide coils, the local concentration of chainlinks is rather high (0.3–0.4 mol/l).

Table 3.6 Local (ρ_{loc}) and average ($\langle\rho\rangle$) density (in a coil volume (mol/1)) of chainlinks of poly-4-vinylpyridine in ethanol [193]. Reproduced by permission of Nauka

MW	β	$\langle\rho\rangle$	ρ_{loc}	ρ_{loc}^{a}
250×10^3	0.2	0.05	0.30	0.40
250×10^3	0.1	—	0.20	0.27
5×10^3	0.2	0.05	0.30	0.36
5×10^3	0.1	—	0.20	0.20

$^a\rho_{loc}^{*}$ is determined in a vitrified solution.

Such a high local concentration of chainlinks, e.g. in a PVCL chain containing voluminous caprolactam rings in 'chair' conformation promotes the formation of apolar areas in which carbonyl groups polarize molecules of water modifying the structural organization of water associates (see section 3.2). Certainly, in the case of poly-*N*-vinylamides of different structure, not containing such volume substituents as a caprolactam ring, local apolar areas are formed in aqueous solution near to a chain, which in comparison to those in PVCL are shorter and have higher permittivity.

Estimation of the amount of chainlinks included in such local apolar areas of PVP or PVCL macromolecules in an aqueous solution was performed using the well-known fluorescent probe, magnesium salt of 1-anilinonaphthalene-8-sulfonic acid (ANS) [161, 162]. The probe molecule forms a complex with 13–15 PVP chainlinks. It was found that the probe–PVCL complex is less hydrated than the probe–PVP complex. The lifetime of the PVCL–probe complex (7–10 ns) also exceeds that of the probe–PVP (<6 ns), determined by the fluorescence method [167].

SOLVENT EFFECT ON PHASE SEPARATION TEMPERATURE OF PVCL

The special nature of the hydration of PVCL macromolecules is distinctly shown in the effect of the nature of the solvent on the temperature of polymer precipitation in comparison with that in pure water [194]. The change of solvent quality was carried out by the addition of reliable PVCL solvents to an aqueous polymer solution.

Interestingly, the introduction of alcohols in aqueous PVCL solution essentially affects the value of $T_{ph.s.}$. Some alcohols at certain contents diminish the solubility of PVCL, which is expressed in a significant fall of $T_{ph.s.}$ in comparison with that in pure water (Fig. 3.21) [194].

If in a series of alcohols the structure of the alkyl radical changes (CH_3OH, C_2H_5OH, *n*-C_3H_7OH, iso-C_3H_7OH or *tert*-C_4H_9OH), one can observe that this change, with alcohol concentration (from 0 to 10–15 mol% (Fig. 3.21)), affects

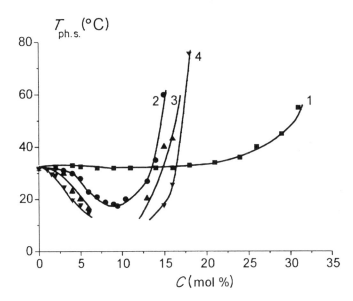

Fig. 3.21. Dependence of phase separation temperature of PVCL on aliphatic ROH (methanol-1, isopropanol-2, propanol-3 and *tert*-butanol-4) concentration in PVCL–water–alcohol system. Reproduced by permission of Nauka.

in different degree the decrease of $T_{ph.s.}$. Methanol virtually does not affect the $T_{ph.s.}$ of PVCL, whereas ethanol and isopropanol lower $T_{ph.s.}$ to 28 °C (at 14 mol% of ethanol) and 17 °C (at 10 mol% of isopropanol). Propanol and *tert*-butanol decreases $T_{ph.s.}$ much more, to 10–12 °C and even lower, and expand considerably the range of alcohol concentration at which the quality of the solvent mixture (water + alcohol) becomes poor towards PVCL.

A number of alcohols is under observation which cause a greater and greater essential decrease of $T_{ph.s.}$:

$$CH_3OH < C_2H_5OH < iso\text{-}C_3H_7OH < tert\text{-}C_4H_9OH < n\text{-}C_3H_7OH$$

These cannot be explained by the decrease of acidity in the ROH group as the strongest effect is displayed for alcohols of low pK_a.

At high alcohol concentration (> 20 mol%) PVCL solubility is essentially improved and $T_{ph.s.}$ increases sharply up to > 60–80 °C.

As well as in case of ethanol [161, 162], various isopropanol concentrations in water cause shrinkage of PVCL coils at concentrations where the greatest reduction of $T_{ph.s.}$ is observed [189].

One could assume that PVCL C=O groups interact directly with HO—R through hydrogen bonds and thus increase alcohol concentration near to the chain. However, the same decrease of $T_{ph.s.}$ of PVCL as in the case of iso-propanol introduction, arises on the addition of polar acetonitrile (CH$_3$≡N)

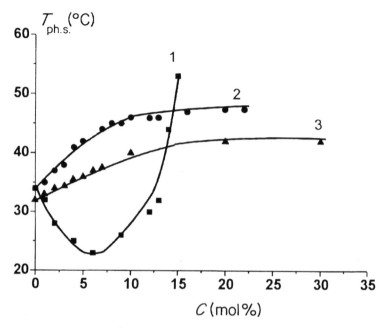

Fig. 3.22.   Dependence of PVCL phase separation temperature ($T_{ph.s.}$) on concentration of acetonitrile (1), ethanolamine (2) and urea (3) in water [189]. Reproduced by permission of Nauka.

whose molecule has no acidic proton and cannot form a hydrogen bond with the C=O group of chainlink (Fig. 3.22).

Molecules of acetonitrile as well as molecules of alcohols enter into the hydrate surroundings of PVCL macromolecules due to interaction with C=O dipole polarized water molecules located in chain-like associates. In this case the functional group of the additive participates in the formation of hydrogen bonds with protons of water molecules near to the PVCL chain, resulting in the organization of different solvate surroundings in which the fraction of water molecules is replaced by molecules of organic solvent.

On the other hand, the hydroxyl group of alcohol molecules, unlike aceto-nitrile, is capable of hydrogen-bond formation with both oxygen atom and protons of polarized water molecules near to a chain. These interactions promote the increase of alcohol concentration in the solvate shell around PVCL macromolecules. The effect of alkyl group length in alcohol on the decrease of $T_{ph.s.}$ is due to the fact that hydrocarbon fragments of R—OH partially screen PVCL chainlinks from water and contribute to apolar interactions between them and hydrocarbon fragments of the main chain, favouring dehydration of chainlinks [38]:

$$(-CH_2-CH-)_n$$

```
(—CH₂—CH—)ₙ
        |
        N ═O···H—O···H—O··H—O···H—O—CH₂—CH₂—N—H····O—H
       ( )          |      |      |                |      |
                    H      H      H                H      H
```

```
(—CH₂—CH—)ₙ
        |
        N ═O···H—O···H—O···H—O
       ( )          |      |      |
                    H      H      H
                                  :
                                  O
                                  ‖
                                  C
                              ╱       ╲
              H—O···H₂N            NH₂···O—H
                 |                       |
                 H                       H
```

As the additional hydrophilic substituent appears in the R—OH alkyl group, e.g., —NH$_2$ in monoethanolamine, the approach of these molecules to the chain occurs because of hydrogen bonds between the R—OH of alcohol and polarized water molecules. In turn, the alcohol molecule with the —NH$_2$ group attracts additional molecules of water into the solvate shell at the expense of amine group hydration. In this case PVCL phase separation temperature rises significantly form 32 °C to 45 °C with the increase of monoethanolamine concentration up to 15–20 mol%.

Urea introduced into a PVCL aqueous solution acts in a similar manner (Fig. 3.22), forming hydrogen bonds to polarized water molecules near to the PVCL chain and, probably, attracting an additional fraction of water molecules to the shell due to the interaction between NH$_2$ of urea and H$_2$O.

Both in the first case (monoethanolamine) and in the second (urea) the hydrate shell expands and a higher temperature ($T_{ph.s.} > 32$ °C) is required to promote the exclusion of water molecules due to hydrophobic, dipole–dipole and dispersion interactions at PVCL–water separation.

These data allow one to offer a concept on denaturation of proteins in aqueous solutions on the introduction of urea. There are some cavities on the protein molecule surface where carbonyls of amide groups could be outward-directed to a solution. These groups could be surrounded by apolar fragments of peptide chain. Like C$=$O groups of poly-*N*-vinylamides investigated, they polarize water associates into chain-like formations where water molecules have a low realization degree of hydrogen bonds. These cavity-localized modified water molecules could promote interaction of protein molecules with the components added in aqueous solution (urea, alcohols, acids, polymers etc.).

Carbonyl groups of urea molecules form hydrogen bonds with these water molecules, while urea molecules are drawn into cavities which could be charaterized as denaturation sites. As a result, urea concentration in the cavities could increase changing peptide functional group surroundings at the expense of the attraction of additional water molecules into the hydrate (solvate) layer of a cavity. The new solvate layer covers greater space compared to that formed in the absence of dissolved urea and reinforces the energy of solvent–molecule interaction with peptide chain fragments in cavities, opening other functional groups capable of hydrogen-bond formation. Hydration of new appearing groups (particularly C$=$O) seems to initiate separation of peptide chains within cavities like 'undoing a zipper'. High hydration energy of these groups promotes this process.

One can clearly show a significant action of hydrophobic voluminous fragment (benzene ring) on dehydration of a PVCL chain in comparison with that of aliphatic alcohols in the example of phenol as an additive [194]. The considerable decrease of PVCL separation temperature in water occurs at a very small phenol concentration, being 100 times less than that of isopropanol (Fig. 3.23).

The fact that phloroglucinol (1,3,5-trihydroxybenzene) molecules affect the PVCL hydrate shell even more strongly than phenol, demonstrate the co-operative interactions of phloroglucinol molecules with polarized water molecules of several polymer chainlinks simultaneously because of the presence of three HO groups in a benzene ring.

The observed phenomena related to PVCL $T_{ph.s.}$ in water at the introduction of organic compounds being good PVCL solvents can be attributed to the change of solvate (hydrate) surroundings of chainlink. Water shell generated nearby PVCL chain in apolar areas due to the polarizing action of C$=$O group contains water molecules of a reduced number of hydrogen bonds. As a result, protons and oxygen atoms of these water molecules have the opportunity to participate in interactions with other added molecule containing groups which act as proton donors ($-$OH, $-$NH$_2$ or $-$COOH) or acceptors (C$=$O, R$-$O$^-$, R$-$COO$^-$, R$-$SO$_3^-$ or $-$C$\equiv$N). There are polarized water molecules in the

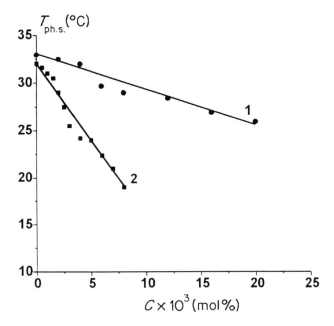

Fig. 3.23. The concentration effect of phenol (1) and phloroglucinol (2) on phase separation temperature ($T_{ph.s.}$) of PVCL in water [189]. Reproduced by permission of Nauka.

shell where the oxygen atom of increased basicity can form stronger hydrogen bonds with protons of, for example, alcohol molecules (propanol, *tert*-butanol or phenols).

Such water molecule behaviour in thread-like associates is predicted from quantum-chemical calculations of charge distribution on oxygen atoms of water molecules joined with the oxygen atom of $C=O$ through the hydrogen bond. Among them certain water molecules located close to $C=O$ have elevated negative charge in comparison with that on the oxygen atom of the bulk water molecule.

The other factor enhancing the heterogeneity of water molecule arrangement and lowering the hydrogen-bond realization degree per molecule (less than three to four), is the apolar areas nearby polymer chain in water. Certainly the dimensions of slightly polar areas and their permittivity depend on the chemical structure of poly-*N*-vinylamides, namely on side residues volume and configurational sequences of main chain regulating conformational state of macromolecule and, in turn, local concentration of chainlinks.

It should be noted that the elevated basicity of oxygen of the water molecule is experimentally displayed by a ¹H NMR method regarding the change of proton chemical shift in the chloroform molecule as a probe present in aqueous solution, on the addition of compounds of various nature (salts of alkyl ammonium or *tert*-butanol) [172, 195]. So, with the increase of alkyl substituent length, ammonium salt induces a marked downfield shift of the $CHCl_3$ proton, indicating the strengthening of the hydrogen bond between $H_2O\cdots H-CCl_3$. In a *tert*-butanol–water mixture an especially large downfield shift of chloroform proton was found achieving its maximum at 6 mol% alcohol in water and indicating hydrogen bond reinforcing between oxygen atom of water molecule changed under additive influence, and chloroform molecule [172]. These two examples seem to give additional proof to the concept of the specific structure of the water associates nearby poly-*N*-vinylamide macromolecules.

On the other hand, the hydrate shell of macromolecules of PVCL, PVP, etc. also contains water molecules with free unequipped protons which can interact with molecules containing proton acceptor groups (COO^-, SO_3^-, I^- and others).

In conclusion let us summarize that the water structure near chainlinks containing polar $C=O$ groups is modified, in comparison to that of bulk water, due to the polarizing action of $C=O$ on water associates situated in slightly polar surroundings organized by apolar fragments of the chain. The dimensions of space created by chain fragments limit the number of water molecules involved in water associates, while dipoles interact with water molecules forming a modified hydrogen bond network which is probably of threadlike or chainlike type.

4

Complex Formation of Macromolecules in Aqueous Solution

The most important property of PVP as a best-known representative of poly-N-vinylamide generation is its capability to interact with various low-molecular-weight and high-molecular-weight compounds in aqueous solutions. This property has motivated the interest of large researchers in PVP and its wide use in various research fields, in engineering and medicine [2, 43, 70, 111, 157, 196].

The other important property of PVP is the ability to form 'soft' complexes in aqueous solution, i.e. complexes with stability constant (K_b) ranging from 1 to 10^4 l/mol. This property is especially significant in developing medicinal preparations including this polymer in combination with other substances. PVP is virtually non-toxic because of its weak interaction with the cell surface and with other biological objects [70, 111, 157, 196].

The application of PVP aqueous salt solution ('Haemodesum'™ containing 6% of PVP) as a detoxication preparation in Russia during the past 30 years has shown its high medical efficiency in the treatment of various diseases and also its harmlessness [43,157]. This polymer does not cause a negative effect on blood elements [70,111].

PVP in aqueous solution forms complexes with a large number of low-molecular-weight and high-molecular-weight compounds: inorganic anions, azodyes, phenols, amino acids, surface active substances (SAS), polymers, proteins and others.

4.1 INORGANIC ANIONS

The interaction of inorganic anions such as Cl^-, Br^-, I^-, CNS^- and NO_2^- with poly-N-vinylamide macromolecules (PVP, PVCL and PVMA) was shown by the ^{127}I, ^{81}Br, ^{35}Cl, ^{41}K, ^{23}Na and ^{14}N NMR method [197].

Table 4.1 Widening of signals of anion and cation nuclei in PVP solution [197]. Reproduced by permission of Nauka

| Ion | Water | | Methanol | |
	Ionic strength	$\dfrac{R_2^i - R_2^o}{R_2^o}$	Ionic strength	$\dfrac{R_2^i - R_2^o}{R_2^o}$
I⁻	1	2.58	0.5	0.3
Br⁻	1	1.13	0.5	0.22
Cl⁻	1	0.49	0.02	0.39
Na⁺	1	0.03	0.02	0.38
K⁺	1	0.02	0.5	0.51

Notes: Measurement conditions: [PVP] 0.1 mol/l, 'Bruker' WM-400, $\bar{M}_w = 20\,000$; R_2^h and R_2^o are ion spin–spin relaxation rates in the presence and absence of PVP.

One could assume that amide groups with a negative charge on oxygen should interact with cations, and direct contact of anions to carbon of the carbonyl group charged positively (Chapter 1) should be complicated by voluminous side substituents, especially in PVCL. The experimental data (Table 4.1) testify that the addition of PVP into aqueous solutions of KI, KBr, KCl or NaCl salts results in a significant widening of signals of anions only, not cations. It specifies severe restriction of mobility of halide ions.

It follows from Table 4.1 that the effect decreases in a series: I > Br > Cl. At the same PVP concentration the average mobility of Na^+ and K^+ cations does not change, confirming the absence of the effect of solution viscosity increase at PVP introduction on the observed widening of anion signals. In methanol (Table 4.1) primary binding of anions with PVP is not observed, while relative signal widening in methanol is being displayed in a much smaller degree than in water [197].

Determination of binding or stability constants (K_b) was performed by the NMR method assuming that polymer binding sites are similar and independent [198–200]. Then

$$K_b = \alpha C_a / (1 - \alpha) C_a (C_p - \alpha C_a)$$ (eq. 4.1)

where α is the fraction of bound small molecules and C_a and C_p are concentrations of small molecules and polymer (monomer chainlinks) respectively.

At small molecules binding with polymer in a certain case of fast molecule transition from bound to free state, the observed spin–spin relaxation rate (R_2^h) of signals of small molecules is equal to

$$R_2^h = \alpha R_2^b + (1 - \alpha) R_2^o$$ (eq. 4.2)

where R_2^o and R_2^b are relaxation rates of small molecules in free and bound states [199].

Solving eq. 4.2 relative to α we get

$$\alpha = (R_2^h - R_2^o)/(R_2^b - R_2^o) \qquad \text{(eq. 4.3)}$$

After eq. 4.3 substitution in eq. 4.1 and simple transformations we find

$$C_p(R_2^b - R_2^o)/(R_2^h - R_2^o) = 1/K_b \,(1 - \alpha) + C_a \qquad \text{(eq. 4.4)}$$

In the case of $\alpha \ll 1$, eq. 4.4 gives a straight line. Measuring its angle of slope one can find $1/(R_2^b - R_2^o)$, and the section of abscissa (Fig. 4.1) gives $1/K_b$. This approach was applied for the determination of K_b depending on solution ionic strength in a series of poly-N-vinylamides and anions (Table 4.2).

In the case of PVP the dependence of complexation constant (K_b) both on the nature of anions and on ionic strength was observed [200]. PVP anion binding enhances in a series: $Cl^- < NO_3^- < Br^- < I^- < CNS^-$ (Table 4.2). This sequence coincides with an indirect estimation of the binding force of anions with PVP [201–203]. It was shown that the binding force of PVP on dodecylammonium or tetraalkylammonium cations in the presence of anions of different type also grows in the same series ($Cl^- < Br^- < I^- < CNS^-$).

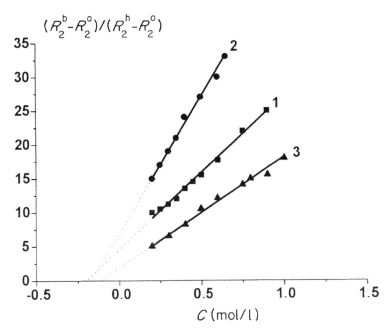

Fig. 4.1. Dependence of $(R_2^b - R_2^o)/(R_2^h - R_2^o)$ on KI concentration in water at 20 °C (ionic strength is 1M) in the presence of PVP (1), PVMA (2), PVCL (3) (background electrolyte = K_2SO_4) [200]. Reproduced by permission of Nauka.

Table 4.2 Stability constant values of PVP with anions in aqueous solution [200]. Reproduced by permission of Nauka

	Stability constant (l/mol)				
Ionic strength	Cl^-	NO_2	Br^-	I^-	CNS^-
0.5	1.04	1.43	1.63	3.09	—
1.0	0.70	1.06	1.11	2.08	11.2
2.0	0.33	0.49	0.55	0.98	5.8

Notes: Ionic strength is K_2SO_4 concentration as K^+ and SO_4^{2-} do not interact to PVP in water; [PVP] = 0.1 mol/l.

Table 4.3 Complexation parameters of poly-*N*-vinylamides with I^- anions [200]. Reproduced by permission of Nauka

Polymer	K_b (l/mol)	r	R_2^b/R_2^o
PVP	3.0	0.76	29.8
PVMA	5.9	0.86	15.2
PVCL	12.5	0.93	33.4

Notes: Here r is the number of bound iodide anions per monomer unit, ionic strength 1 mol/l, 26 °C, [polymer] = 0.1 mol/l, $\bar{M}_w = 15\,000$.

NMR findings also give important information about anion mobility near to a chain [197]. It indicates the loss of anion mobility in a polymer-bound state. Therefore, the ratio of spin–spin relaxation (R_2^b/R_2^o) in bound and free states is: Cl^- 13.0, NO_2^- 12.4, Br^- 15.4 and I^- 25.2.

Further, in a series of poly-*N*-vinylamides, one can follow the effect of substituent structure on polymer interaction with I^- ions (Table 4.3). As seen in the table, PVCL compared to PVP and PVMA possesses the greatest ability to bind with I^- ions.

As the values of r (number of bound I^- anions per monomer chainlink) are rather high, the increase of local ion concentration of iodide near to macromolecules occurs. Such an increase in I^- concentration is important for the preparation of aqueous PVP–I_3^- solutions in antiseptic preparations [76].

4.2 COMPLEXES OF POLY-*N*-VINYLAMIDE–I_3^-

A large number of investigations [2, 76, 204–208] are devoted to the PVP–iodine complex. It is manufactured under the 'Povidone-iodine' trade mark in the USA and other countries' pharmacopoeias, and under the 'Iodovidonium'

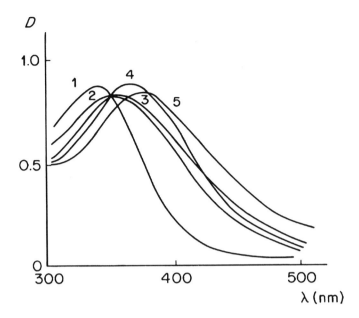

Fig. 4.2. The effect of polymers of various structure on the spectrum of I_3^- polymer complexes: without polymer-1, PVP-2, PVMP-3, PVCL-4 and CPVCL-VP (50 mol%)-5. Measurement conditions: 20 °C, 0.1 cm Cuvette, $[I_2]$ –3.75 × 10^{-4} mol/l, [KI]–1.5 × 10^{-1} mol/l and [polymer]-2.5 × 10^{-2} mol/l [203]. Reproduced by permission of Nauka.

trade mark in Russia, is used widely in the production of a large number of medicinal formulations for external application (solutions, ointment, shampoos and other forms) [2, 41, 74, 199].

The conditions of formation of this complex, its structure and the nature of stability in aqueous solutions are of great interest to researchers. As this complex is formed of PVP and triiodide, let us consider the peculiarities of iodine interaction with iodide anions in aqueous solution.

The introduction of KI in iodine aqueous solution, as is well known [206, 207], raises the solubility of iodine in water due to formation of a KI_3 complex. In other words, there is an equilibrium in water with a certain dissociation constant $(1/K_b)$.

$$I_2 + I^- \leftrightarrow I_3^-, \quad K_b = [I_3^-]/[I_2] \cdot [I^-]$$

The changes of UV spectra of iodine solutions are significant when KI is introduced into the solution, manifesting themselves in the increase of the extinction coefficient (ε) of the I_3^- complex in comparison with that of iodine in water. In fact, the values of ε absorption at $\lambda = 288$, 353 and 460 nm are

$38.9 \times 10^3, 26.4 \times 10^3$ and 975 l/mol cm of the I_3^- complex against 83.5, 16.1 and 74.6 l/mol cm of I_2 at the same wavelengths [206, 207].

Stability constants of I_3^- complexes in water determined by the spectroscopic method fall with the increase of temperature: 840 ± 7 l/mol at 16 °C, 690 ± 2 at 25 °C, 530 ± 3 at 36 °C, 440 ± 5 at 44.7 °C [207].

One can observe two absorption maxima at $\lambda_{max} = 290$ and 350 nm in the UV spectrum of the well-known Lugole solution containing 1 g of I_2 and 2 g of KI in 200 ml H_2O [206]. However, the optical density of absorption of this solution which is often used by researchers in the study of the I_3^- complex is greatly increased at the further increase of I^- concentration in solution, as the equilibrium is shifted to the I_3^- formation. At an insignificant mole excess of [KI] over [I_2] in the solution, this equilibrium considerably complicates the exact determination of both ε of the PVP–I_3^- complex and the value of the stability constant.

The introduction of PVP promotes a further increase of absorption intensity of the Lugole solution, especially in the wavelength range $\lambda = 400$–500 nm [206]. The general increase of absorption intensity in the range $\lambda = 290$–400 nm is due to PVP complex formation at the low given ratio of [KI]/[I_2] (2 : 1 by weight). However, further introduction of KI to the solution causes a further increase of absorption intensity, making the ε estimation of PVP–I_3^- complexes difficult.

Therefore, only at a high mole ratio of [KI]/[I_2] = 40 can the absorption intensity of the PVP–I_3^- complex be determined [203] when further introduction of KI does not change the UV spectrum of the solution containing polymer and iodine. It was established that the polymers investigated only slightly change the extinction coefficient of the polymer–I_3^- complex, but affect the absorption maximum. The ε value of pure I_3^- is 26.4×10^3 l/mol cm at $\lambda = 353$ nm, that of I_3^- in the presence of poly-*N*-vinylamides added in the solution is 24 000–27 000 l/mol cm. However, the absorption maximum (λ_{max}) of a polymer complex depends on polymer structure and shifts from normal 353 nm to 364, 372 and 367 nm on the introduction of PVP, PVCL and PVMA respectively, with a general increase of absorption at a wavelength equal to 400–500 nm [208].

These significant changes in the I_3^- UV spectrum at triiodide anions surrounded by polymer chainlinks ([Polymer]/[I_2] = 10) testify to a different hydrate environment of anions near to chains of researched polymers compared to that of free I_3^-.

Triiodide ions surroundings near to a polymer chain can be varied [203] by the application of copolymers of VP and VCL (CP–VCL–VP) of various composition (Fig. 4.3). So, in the presence of CP–VCL–VP with 50 mol% of VCL one can observe a further wavelength shift of the absorption maximum of the copolymer–I_3^- complex to 378 nm in comparison to $\lambda_{max} = 363$ nm for PVP and 373 nm for PVCL (Fig. 4.3). The stability constant value of the PVCL–triiodide complex was assessed by spectroscopic method to be about 10^4 l/mol [203]. The correct determination of the PVP–I_3^- stability constant in water is

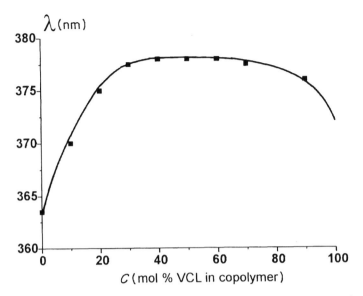

Fig. 4.3. The effect of VP–VCL copolymer composition on the absorption maximum shift of I_3^-–polymer complex [203]. Reproduced by permission of Nauka.

complicated by this method due to the closeness of the values of λ_{max} for the PVP–I_3^- complexes and KI_3^-.

The fact that the shift of λ_{max} of the chromophore triiodide anion to a long-wave field is significant even at the addition of a small amount of N-methylpyrrolidone as a low-molecular-weight analogue of a PVP chainlink to an aqueous I_3^- solution, seem to confirm the effect of water molecules on the I_3^- electronic state polarized by analogue amide groups. The value of λ_{max} is increased sharply from 353 to 364 nm when the N-methylpyrrolidone concentration in water grows insiginificantly from 0 to 18 mol% [203]. But at the increase of analogue concentration from 20 to 95 mol% λ_{max} increases only up to 368 nm.

It is possible that at the introduction of a small amount of analogue into water the structure of the water associates is changed, raising the acidity of some water molecules and, in turn, promoting the stronger hydrogen-bond formation between water molecules and triiodide anions. Earlier it was found by a NMR method [174] that water molecule acidity increases on the introduction of a small amount of such substances as *tert*-butanol or organic ammonium cations.

Therefore, significant changes of λ_{max} of I_3^- in aqueous polymer solutions compared those of I_3^- in pure water are probably caused by different water surroundings of the anion near to a chain. Polarized by C=O water molecules participate in complexation with a triiodide molecule through hydrogen bonds.

As the hydrogen-bond network of PVP, PVCL or PVMA can be different, water surroundings near to their chains and the anions affect the electronic state of the triiodide anion in various ways.

$$(-CH_2-CH-)_n$$

N—C=O···H—O···H—O···H—O

Let us consider the characteristics of PVP hydration in an aqueous solution in the presence of iodide and triiodide ions. It is remarkable [209] that the increase of I⁻ concentration in water brings about the displacement of chemical shifts in ¹³C NMR spectrum of carbon atoms C=O and —C$_x$H— in the main PVP chain (Fig. 4.4). Attention should be paid to the fact that on the increase of iodide concentration in a solution the chemical shift of the C=O group is shifted upfield and in addition the C=O signal half-width narrows. Such an effect, as shown in ref. [209], is caused by a partial dehydration of the C=O group.

Iodide ions falling within the water shell of the chain interact with polarized molecules located close to C=O, replacing some H₂O molecules from the

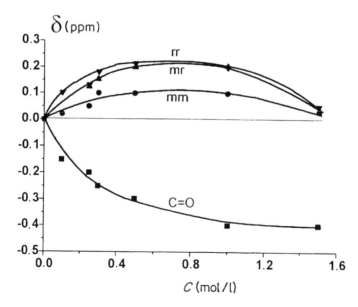

Fig. 4.4. The effect of KI concentration on ¹³C chemical shift of carbon atoms in C=O and —C$_x$H— groups (mm-, mr- and rr-triads) of PVP in aqueous solution [204]. Reproduced by permission of Nauka.

hydrate environment of amide groups. Comparing the values of chemical upfield shifts of carbon atom in the $C=O$ group (Fig. 4.4) on the introduction of 1 mol/l of KI and at a change of the [PVP]/[H_2O] concentration ratio (see Fig. 3.6) one can evaluate the number of water molecules replaced by the iodide ion. So, about four or five molecules of water are substituted by iodide ion per 13–14 molecules of water surrounding one chainlink. As a result the I^- concentration near to the chain rises in comparison with that of I^- in solution.

At the same time the signals of methine carbon (in syndio-, iso- and hetero-triads) are shifted in another direction, namely downfield (Fig. 4.4), indicating the partial dehydration of chainlinks accompanied by conformational transformations of the chain [209].

As for H_2O molecules involved in the PVP–I_3^- complex surroundings, they become less mobile than in a pure solution or in an aqueous PVP solution. In fact, the introduction of both PVP (1.8 mol% of PVP) ($R^1 = 2.47$ s^{-1}) and PVP–I_3^- ($R_1 = 3.1$ s^{-1} with 5 wt% of I_2 relative to [PVP]) increases the spin––lattice relaxation rate (R_1) of ^1H nuclei in pure water ($R_1 = 1.43$ s^{-1}).

The PVP effect of the R_1 increase is explained by water molecule mobility restrictions in hydrated shell at the expense of hydrogen-bond formation between $-C=O\cdots H-O-H$ and polarized water molecules [210, 211]. The strength of hydrogen bonds between polarized water molecules and the molecules with $C=O$ groups near to the chain, and also the fraction of water molecules participating in bond formation are apparently the major factors affecting the mobility of water molecules in a hydrate shell.

The introduction of I_3^- cause a further increase in R_1. It is interesting that the disappearance of triiodide ions from the solution (under thiosulfate reaction) changes the relaxation performance of the latter back to that of a triiodide-free PVP solution.

Thus the mobility of water molecules being in local sites of 12–14 chainlinks involving the I_3^- anion is somewhat restrained in comparison with that in a solution. Slowing down of mobility seems to be caused not only by the interaction of water molecules with $C=O$ groups but also by their interaction with I_3^- [208].

In the absence of PVP the spin–lattice relaxation rate of D_2O remains constant at I^- concentration change in a solution. Neither iodide nor triiodide form associates with D_2O. They arise only near to the PVP chain because of the presence of polarized D_2O molecules. Otherwise the I_3^- anion should have interacted with the negative end of a $C=O$ dipole, which is impossible.

The free I_2 molecule can only interact with an oxygen atom of the $C=O$ group of PVP in organic solvent. In an example of $I_2 + N$-ethylpyrrolidone reaction [206] it was shown that an iodine molecule in organic solvent forms a complex with $C=O$ of amide group. This complex reaction occurs in CCl_4 through interaction of I_2 with the negative end of the $C=O$ dipole.

The IR spectra of N-ethylpyrrolidone in the absence and in the presence of various amounts of I_2 [206] testify that the absorption maximum at

$v = 1690 \text{ cm}^{-1}$ (the band of stretching vibrations of C=O) at iodine introduction starts to decrease gradually with the occurrence of the second maximum at $v = 1650 \text{ cm}^{-1}$.

It is remarkable that the same effect is observed in *N*-ethylpyrrolidone on the addition of D_2O [60]. On the increase of the mole portion of D_2O in *N*-ethylpyrrolidone $v_{C=O}$ is displaced from 1690 cm^{-1} (pure substance) to 1637 cm^{-1} (mole portion of water equal to 0.95). In the second case the C=O frequency shift is caused by hydrogen-bond formation with water molecule where the C=O group is a donor of electrons and the proton of water is an acceptor.

The identical shift of $v_{C=O}$ in the direction of low frequencies in the presence of water molecules and I_2 shows that the I_2 molecule is an acceptor of electronic density. The value of the stability constant (K_b) of *N*-ethylpyrrolidone + I_2 in CCl_4 is 9–10 [206].

One can assume that PVP interacts in the same manner with I_2 only in organic solvent (e.g. in CH_2Cl_2). However, in aqueous solution water molecules do not permit I_2 molecules to interact with C=O groups of the chain due to stronger hydrogen bonds between C=O ⋯H—O—H than that of C=O ⋯I—I. Iodine interaction with O=C— becomes essential in the preparation of the solid state PVP–I_3^- complex as a result of PVP powder and solid I_2 mixing in the pharmaceutical industry.

At PVP complex formation with I_3^- in water the segmental mobility of polymer chains also changes. The spin–spin relaxation time (T_2) of protic PVP groups decreases at the complexation by 2–2.5 times (Table 4.4).

The effect of PVP interaction with I_3^- in water is distinctly reflected on the form of PVP carbon atom signals in the ^{13}C NMR spectrum (Fig. 4.5). In PVP–I_3^- aqueous solution the value of $\Delta v_{2/3}$ (the signal width at the height of 2/3 maximum amplitude) of carbon signals of the $—CP_{(2)}H_2—$ and $—C_{(3)}H_2—$ groups of the pyrrolidone ring significantly exceeds that of the same signals in the absence of triiodide. The value of activation enthalpy of segmental movements in the absence and in the presence of I_3^- found from the slope of the dependence straight line log $\Delta v_{2/3}$ on $1/T$ of these nuclei is 9.2 (in the absence of I_3^-) and 6.4 → 26.8 (for $—C_{(3)}H_2—$) and 36.8 (for $—C_{(2)}H_2—$) kJ/mol, in the presence of I_3^- ions respectively. It testifies to the significant restriction of

Table 4.4 Spin–spin relaxation time (R_1) of protic PVP groups in water and in PVP–I_3^- aqueous solution [209]; [PVP] = 1.8 mol%, [I_2] = 5 wt% relative to [PVP]. Reproduced by permission of Nauka

Composition	D_2O	PVP + D_2O	PVP–I_3^- + D_2O
$R_1 (s^{-1})$	1.43	2.47	3.10

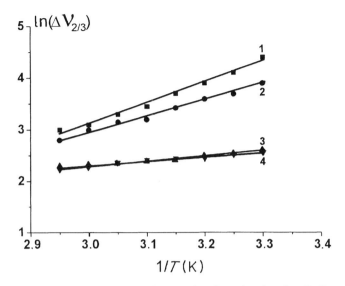

Fig. 4.5. Dependence of the value of $\Delta v_{2/3}$ of carbon signals of $-C_{(2)}H_2-$ (1,3) and $-C_{(3)}H_2-$ (2, 4) groups in PVP on aqueous solutions temperature of PVP–I_3^- (1,2) and PVP (3,4) [212]. Reproduced by permission of Nauka.

mobility of the whole pyrrolidone ring in the PVP–I_3^- complex in comparison to that of PVP in pure water [207].

PVP interaction with I_3^- affects in a complex manner the form and width of the C=O signal (Fig. 4.6). In water the C=O signal form is distinctly quadruplet, becoming more narrow and shifting upfield to 0.15 ppm on the addition of I_3^-. The introduction of thiosulfate reacting with triiodide ions forces the signal to regain its initial position and, partially, its initial form.

It is worth noting that the quadruplet form of the C=O signal and its large width in aqueous PVP solution is probably determined by an unequal hydration degree of chainlinks being in different configurational sequences of the main chain (Fig. 4.3). The signals of C=O of PVP in different solvents are represented in Fig. 4.3 for comparison, namely in methanol, ethanol and chloroform. The strength of the hydrogen bond in that series is gradually reduced along with the reduction of signal width (2–2.5 times).

This width reduction trend of the C=O signal is observed at the formation of PVP–I_3^- complexes in comparison with pure PVP. It once again confirms that the fact that I_3^- ions interacting with water molecules most close to chain (to C=O groups) replace a fraction of H_2O molecules from the chain hydrate shell. That causes the upfield shift of the C=O signal, reducing the total signal to half-width. Conformational transformations in PVP macromolecule occur at its interaction with I_3^-, accompanied by a significant reduction of specific viscosity

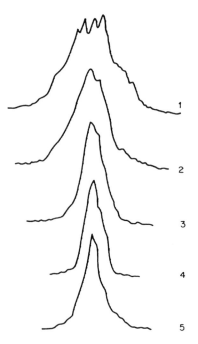

Fig. 4.6. Form of C=O signal of PVP at 50 °C in water (1), in water with I_3^- (2), in methanol (3), in ethanol (4) and in chloroform (5) (Bruker WH-360) [212]. Reproduced by permission of Nauka.

of macromolecules, indicating the shrinkage of coils and the replacement of a fraction of water molecules.

Considering the above-mentioned data one can state that each complex consists of the triiodide ion and some polarized molecules of water surrounding the ion and interacting with the ion through hydrogen bonds. The triiodide ion is likely to be surrounded by two shells: the nearest one consisting of water molecules and the remote one consisting of polymer chainlinks. The number of chainlinks forming the remote shell is 12–15 [166, 208, 209].

4.3 LOW-MOLECULAR-WEIGHT ORGANIC SUBSTANCES

AROMATIC COMPOUNDS OF PHENOL TYPE

A large group of low-molecular-weight organic compounds forms complexes with poly-*N*-vinylamides in water. This group involves aromatic compounds with various substituents such as hydroxyl, carboxylic or sulfonate groups, surface-active substances (SAS), organic dyes, bearing a negative charge on sulfonate group and so on [213–223].

The determination of binding (or stability) constant (K_b) for organic compounds by poly-N-vinylamides, in particular PVP have performed by a equilibrium dialysis method with use of the Klotz equation [213]:

$$1/r = 1/(nK_b a) + 1/n \qquad \text{(eq. 4.5)}$$

where r and n are the number of bound moles of a low-molecular-weight substance and the number of moles of binding sites on the mole of PVP chainlinks, a is the mole concentration of unbound molecules of this compound in a solution and K_b is the stability constant calculated on the mole of chainlinks. The linear ratio between $1/r$ and $1/a$ was established for systems of PVP–phenols, and the value of $1/n = 1$ [213]. It means that one molecule of phenol interacts with one PVP chainlink.

In this case when the comparison of stability constants (K'_b) for PVP and biological macromolecules, in particular bovine serum albumin with an average number of chainlinks being 900 is required, the calculation of r is performed on a mole of total protein, and in the case of PVP, on 10^5 g of polymer [215]. At the account of K_b per base mole (monomer unit) of PVP experimental value of K'_b is necessary to divide on 900.

Hereafter, for elucidation of the special features in complex formation of poly-N-vinylamides with various low molecular weight substances in water, the value of K_b is calculated per monomer chainlink.

A large group of phenol derivatives interact with PVP macromolecules in water, as is shown by an equilibrium dialysis method [213–215]. The K_b value depends on the pK_a of the hydroxyl group in the benzene ring increasing from 2.5 for p-methoxyphenol (MPH), 4.0 for phenol (PH), 9.0 for p-bromphenol (BPH) to 10 l/mol for p-nitrophenol (NPH) [208]

The introduction of carboxylic and hydroxyl groups in benzene ring favours the complex formation with PVP. Therefore, the binding constant of salicylic (SA) and p-hydroxysalicylic (HSA) acids by PVP are 18 and 49 l/mol respectively, at pH 2.2, indicating a prominent role for the hydroxyl group in complex formation of these compounds with PVP.

The contribution of carboxylic groups being in ionized or protonated forms at the PVP complexation distinctly is followed by comparison with K_b for benzoic (BA), salicylic (SA) and *p*-hydroxysalicylic (HSA) acids in solutions at different pH [216]. It is clear that the protonated form of the carboxyl group is preferable to the ionized one because of its binding by PVP macromolecules: the K_b values in the case of BA, SA and HSA are equal to 8.7 (at pH 3.4), 2.8 (at pH 5.0), 10.3 (at pH 2.2), 14.5 (at pH 3.8), 49 (at pH 2.2) and 31.5 (at pH 4.2) [216]. Indirect evidence concerning a role of positively charged groups in the complex formation can be seen in an example of pyridine derivatives with the carboxylic group (nicotinic (NA) and isonicotinic (INA)) acids: in an aqueous solution at pH 5.6 when a part of the pyridine ring is protonated the K_b values are low being 1.2 (NA) and 1.65 (INA) [211].

COOH   COOH

NA   INA

Consequently, the findings demonstrate a significant contribution of proton-containing groups (—OH and —COOH) at the interaction with PVP chains supporting the concept concerning the participation of polarized water molecules through hydrogen bonds at complex formation between polyamides with low-molecular-weight compounds of the phenol type.

DYES WITH SULFONATE GROUPS

Appreciable ability to interact with PVP macromolecules in water displays organic dye molecules, bearing negative (sulfonate) charges [214]. A dye molecule containing a positive charge (azobenzene trimethylammonium chloride [218]) almost does not interact with the polymer.

Molecules of dyes from a group of azodyes, for example orange dyes, form complexes with PVP [21].

$$R_2N-\text{C}_6\text{H}_4-N{=}N-\text{C}_6\text{H}_4-SO_3^-$$

where R is methyl, ethyl, propyl or butyl.

It should be noted that in a series of orange dyes where the substituent structure changes from methyl, ethyl, propyl to butyl, the contribution due to increasing carbon atoms in the alkyl apolar group in the complexation is marked: the K_b value grows only 2.5 times in going from methyl (22 l/mol) to butyl (57 l/mol) substituent [219, 220]. However, one can assume that a predominant contribution in K_b is stipulated by the interaction of a sulphonate-containing benzene ring with a PVP chain due to hydrogen bonds between ionized sulfonate groups and water molecules of the hydrated shell around macromolecules.

Some information about dye molecule surrounding in a sorption PVP site can be obtained from comparison of spectral characteristics for the dye molecule in water, albumin site and PVP site [215]. In fact, absorption maximum of the dye molecule being in water, albumin and PVP is at λ_{max} = 465, 435 and 470 nm respectively. The observed exceeding of λ_{max} for a dye molecule in a PVP site and in water seems to demonstrate the more polar local environment of this molecule spaced in a PVP hydrated shell compared to that in water, and of course in a hydrophobic site of albumin.

It is of interest that the nature of the salt affects the binding constant of these dye molecules [219, 220]. The special effect on PVP complex formation is caused by an addition in an aqueous solution of (1 mol/l) thiocyanate sodium (NaCSN) against chloride sodium. There is an increase of K_b for orange dye with propyl (88 l/mol) and butyl substituents (162 l/mol) unlike methyl (23 l/mol) and ethyl (36 l/mol) substituents.

Such an increase of binding constants for these orange dyes by PVP macromolecules in the presence of thiocyanate anions becomes clear if we recall that these anions are bound by the macromolecules to a greater degree than those of chloride anions (see section 4.1). The thiocyanate anions, probably, as well as the iodide or triiodide anions, replace water molecules from the hydrated shell near to the chain to influence the strengthening of hydrophobic interactions between apolar fragments of the dye molecule and chainlinks.

In a greater degree PVP macromolecules bind the dye molecules containing hydroxyl and sulfonate groups, for example Tirone with the K_b = 167 l/mol, as determined by the equilibrium dialysis method [221].

Tirone	Bengal rose

The Bengal rose molecule itself represents a voluminous molecule containing two hydroxyl groups of the phenol type and a carboxylic group and is bound by PVP with a high binding constant being 2.8×10^3 l/mol [221].

It was observed by a spectroscopic method that at a complex formation the extinction coefficient of Bengal rose dye molecules with PVP grows significantly, from 1.1×10^4 (without PVP) up to 3.18×10^4 l/mol cm at λ = 560 nm. It is possible that molecules of PVP destroy dye associates in water and form complexes, each of which involves one dye molecule per sorption site organized from 12–15 chainlinks.

FLUORESCENT DYES

Let us consider spectral shifts of dye molecules at complex formation with PVP on an example of a well-known fluorescence dye, magnesium salt 8-anilinnaphthalene-1-sulfonic acid (ANS). This fluorescence dye is widely used as a probe for studying the structure of biological systems, namely proteins, enzymes, membranes, etc. [222–225]. The investigations are based on the sensitivity of fluorescent properties of this molecule to its surroundings, being the probe on apolar areas of macromolecules in an aqueous solution [224].

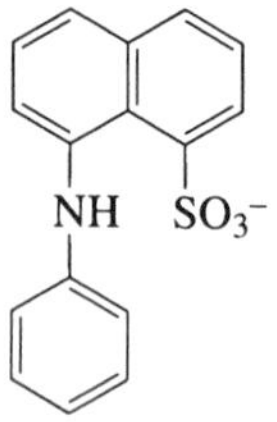

ANS

Anufrieva and co-workers [225] offer a new interpretation of molecular mechanism of an increased fluorescence intensity in organic solvent or in aqueous polymer solutions. The greatest value of fluorescence intensity (J_{FL}) of ANS is achieved at going from water to organic solvent (more than 100 times). In other words, the factors favourable for ANS fluorescence quenching occur in water. Therefore, the duration of ANS fluorescence decreases from 10 ns in organic solvent to 0.6 ns in water. It was established in ref. [225] that the major factor resulting in ANS fluorescence quenching in water is dimerization of ANS molecules in water at the expense of both hydrophobic interactions between aromatic rings and two hydrogen bonds between —N—H and SO_3— groups. The structure of the dimer can be presented as follows:

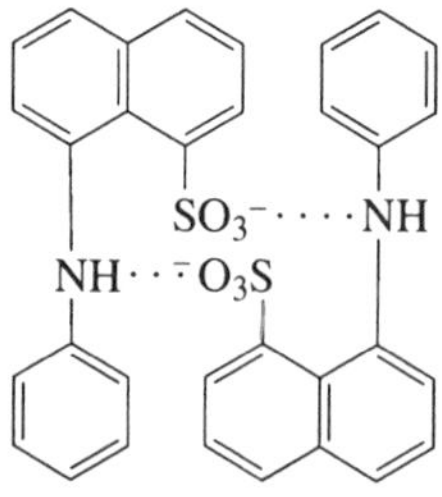

The destruction of the dimer, for example by substitution of hydrogen bonds inside the dimer on hydrogen bonds with molecules of organic solvent (dioxane,

Table 4.5 The binding constants of ANS by poly-N-vinylamides at various temperatures [166]. Reproduced by permission of Nauka.

Polymer	K_b (l/mol)				
	18 °C	12 °C	27 °C	30 °C	33 °C
PVMA	220	170		150	
PVP	360	270	240	220	200
PVCL	630	670		780	970
CP–VP–VCL (10 mol%)	370	300		260	240
CP–VP–VCL (30 mol%)	390	320		280	260
CP–VP–VCL (80 mol%)	490	470		480	480

DMF, alcohols, etc.) results in an increase of concentration of separate ANS molecules which have a large fluorescence duration. Therefore, the ANS intensity increase in an aqueous solution with polymers means the occurrence of separate probe molecules at complex formation.

The fluorescent ANS probe, as was shown [166, 167, 226, 227], forms complexes with PVP and other poly-N-vinylamides in an aqueous solution. Introduction of polymer in a solution with ANS greatly increases the fluorescence intensity of the probe in comparison with that of ANS in pure water. In fact, PVP at a large excess in solution towards ANS enhances J_{FL} of ANS about 70 times, whereas PVCL gives the increase of J_{FL} up to 140 times. In addition, the fluorescence maximum of ANS is shifted from $\lambda_{max} = 510$ nm (the probe in water) to $\lambda_{max} = 480$ and 470 nm in the presence of PVP and PVCL [225–226].

From findings concerning the ANS fluorescence intensity at varying polymer concentration and constant probe concentration in water it was determined the stability constants of poly-N-vinylamide + ANS complexes (Table 4.5). It follows from data in Table 4.5 that the PVCL macromolecules have a binding constant two to four times greater with ANS than with PVP and PVMA. It is remarkable that the value of K_b for PVCL unlike PVP or PVMA grows with increasing temperature from $K_b = 630$ (at 18 °C) to 970 l/mol (at 33 °C), whereas in the case of PVP the K_b value falls from 360 (at 18 °C) to 200 l/mol (at 33 °C). It is of interest that it is possible to find the polymer from copolymers of VCL–VP (80 mol% of VCL) which forms a temperature-independent ANS–copolymer complex.

Such unusual behaviour of PVCL macromolecules is closely connected with conformational transformations of chain happening on solution heating (see Chapter 3). The understanding of the molecular mechanism of ANS fluorescence increases in an aqueous solution in the presence of the poly-N-vinylamides investigated, and also factors defining conformational and solvation transformations in these polymer chains allow one to interpret the nature of interaction of ANS molecules with macromolecules of the polymers investigated.

The fundamental reason for the J_{FL} increase of ANS at complex formation with PVP (or PVCL) in water is due to disintegration of the ANS dimer and transition of separate molecules in close proximity to chain. The disintegration of the dimer seems to occur under the action of polarized molecules of water which interact through hydrogen bonds with both SO_3^- and NH groups of the ANS molecule. The molecules of water themselves represent binding "bridges" between an ANS molecule and the 12–14 chainlinks forming a sorption site on chain [166, 167, 226, 227].

It is necessary to pay attention once more to a significant shift of fluorescence intensity maximum of ANS (in pure water λ_{max} = 510 nm) on introducing PVCL (λ_{max} = 470 nm) in comparison with that of PVP (λ_{max} = 480 nm), indicating a different electronic state of a ANS molecule being in water, PVCL or PVP complexes. The observed difference of the probe fluorescence maximum in PVCL + ANS complexes compared to that of PVP + ANS complexes is likely to testify to an existence of unlike types of water molecules (polarized by the C=O dipoles) surrounding an ANS molecule in PVCL and PVP chains.

The ability of macromolecules to form complexes with ANS depends on molecular weight (MW) of polymer in a range of low MW [191, 226, 227] remaining virtually constant for narrow fractions of PVCL with MW, from 5×10^3 to 1×10^6, and lowering appreciably with decreasing MW, from 5×10^3 to 10^3.

A similar effect of MW for PVP on complex formation of PVP with a fluorescent probe of other structure, sodium salt of 2-(*p*-toluidine) naphthalene-6-sulfonic acid (TNS), was found [228]. This fluorophor also sharply increases the intensity of fluorescence on the introduction of PVP in a TNS aqueous solution with increasing PVP concentration. The growth is identical for PVP with MWs of 4×10^4 and 3.6×10^5. Appreciably smaller the growth of TNS fluorescence intensity on increasing the probe concentration is observed in the case of PVP with $M_w = 10^4$ because of the presence of a significant portion of low-molecular-weight fractions (MW $< 5 \times 10^3$) weakly interacting with the TNS molecule.

The increase in capability of complex formation for small molecules and PVP macromolecules with enhancement of chain length or degree of polymerization P in ranging from 10 to 50, are caused by two factors [191]. The first factor is a rise in the number of contacts between a small particle and by chainlinks of a chain. At going from P = 10 to 50 chainlinks in a macromolecule, the local chainlink concentration (ρ_{loc}) near to a chain enhances collisions of a small molecule with chainlinks of a chain. The point is that it is very important for the interaction of a chain with a small molecule that it has a chainlink density not in the volume of a coil but near to a separate chainlink (ρ_{loc}).

It is shown [229] by a method by fluorescence quenchings that ρ_{loc} for a flexible-chain macromolecule of poly-4-vinylpyridine in a good solvent changes insignificantly in a wide range of P from 70 to 2800 and sometimes exceeds the average chainlink concentration in coil volume. In the same range of P the interaction of macromolecule (type PVP, PVMA, PVCL) with small molecules (type ANS, TNS, I_3^-, etc.) does not change, as in the complex formation reaction separate sites participate consisting of 12–15 chainlinks. The reduction of $P(< 50)$ results in the decrease of ρ_{loc}, and accordingly the capability of macromolecule to complex formation with a small molecule in an aqueous solution.

The second factor is progressive partial dehydration of chainlinks with increasing P at going from an oligomer to a polymer as the accumulation of apolar groups on a chain occurs. The main polyethylene chain and closely spaced chainlinks promote the replacement of water molecules near to the chain and the small molecule surrounded with these chainlinks.

An illustration of this factor is the increase of PVCL phase separation temperature in water at the lowering of a number M_n of PVCL fractions [208]. A considerable rise of $T_{ph.s.}$ for PVCL fractions is found with $M_n < 5 \times 10^3$. In other words, the smaller M_n, the greater $T_{ph.s.}$ of PVCL, and accordingly the weaker sorption capability of sites to small molecules.

PVP forms complexes with fluorescent dyes from a series of water-soluble stilbene Blancphor dyes [230]. Its solubility in water is due to the presence of two sulfonate groups on a molecule.

For PVP with $M_w = 30 \times 10^3$ it was determined that the concentration ratio between the dye (Blancphor BA) and polymer at which the spectral fluorescence changes of dye stops, indicating the complete absorption of dye molecules by PVP macromolecules. As a result it is possible to calculate a number of chainlinks (N) per on dye molecule in the complex.

It is important that the calculated N value of chainlinks involved in a sorption site binding the Blancphore dye is 13–15 PVP chainlinks, corresponding to the value of N found in the case of ANS + PVP complexes [208].

The other peculiarity of a PVP complexation with these dyes is in that the sorption sites with various K_b on macromolecules are formed as an increase of the number of bound dye molecules on chain occurs [230]. Therefore, the change of dye concentration in solution from 1 g/100 ml to 3 g/100 ml causes a considerable decrease in the stability constant (K_b) from 4×10^3 to 3×10^2 l/mol. PVP macromolecules interacting with Blancphor dye molecules are transformed to polyelectrolyte molecules with negative charges. Therefore the involvement of additive dye molecules into PVP complexes becomes difficult due to electrostatic repulsion of negatively charged dye molecules by sites on the chain-bearing sulfonate groups. The PVP macromolecule with filling sorption sites is unfolded to increase specific viscosity.

The introduction of salt in solution PVP + Blancphore dye causes a decrease of electrostatic interaction between charged sites and dye molecules, creating absorption sites with identical complexing capability to the molecules. At [NaCl] = 4.3×10^{-2} mol/l the K_b value is 5×10^3 l/mol becoming independent of dye concentration [230].

Bilirubin

The high value of K_b is characterized by the complex of PVP and bilirubin (toxic product of erythrocyte destruction) [231]. A specrtrophotometric method was used for determination of this constant being $(3.6 \pm 2.4) \times 10^4$ l/mol, and some exceeding this in sites in transport protein of plasma (serum ovalbumin). Some sites in ovalbumin are capable of binding to bilirubin with $K_b = 0.17 \times 10^4$ l/mol [232].

4.4 SURFACE-ACTIVE SUBSTANCES

POLY-N-VINYLPYRROLIDONE

From data of equilibrium dialysis, conductometry and viscosimetry of PVP + sodium dodecylsulfate (DDS) system the following picture of their interaction in a aqueous solution is proposed [233, 234]. Some peculiarities of the DDS micelle formation are displayed, depending on both DDS and PVP concentrations in water.

At small DDS concentrations ($< 5 \times 10^{-3}$ mol/l) in the presence of an appreciable amount of PVP ($(15 \pm 70)\ 10^{-3}$ mol/l) the interaction of DDS molecules with macromolecules does not occur. Then at a further increase of DDS concentration PVP macromolecules begin to prevent the micelle formation of DDS and form DDS + PVP associates involving $[DDS]^n[PVP]^{1/m}$, where n is the number of DDS molecules in one associate, and m the number of chainlinks which induces formation of an associate). The number n can change in a range from 1 to N_{ccm} ($N_{ccm} \approx 60$ is a number of DDS molecules in a regular micelle at a critical concentration of micelleformation (CCM). It is established that the value of $1/m$ is less than unity and is the relation of the number of bound DDS molecules per mole of PVP chainlinks. The $1/m$ values fall between 0.01 and 0.3 established by the method of equilibrum dialysis. In other words, at interaction with DDS some (3–10) chainlinks of PVP chain can participate in the associate formation, depending on DDS concentration. Therefore, at DDS concentration close to CCM, three to four units of PVP attract one DDS molecule.

As the mole ratio of [DDS]/[PVP] increases in a solution (at concentration of DDC in water close to critical concentration micelleformation, namely 8×10^{-3} mol/l), associates from molecules of the detergent are formed, causing a increase of relative viscosity of a polymer solution. These associates from DDC as follows from refs [233, 234], seem not to be true micelles. The PVP macromolecules carrying DDS associates convert to polyelectrolyte molecules.

At DDS concentrations exceeding the CCM in a solution, there are in equilibrium the associates on a polymer chain (true micelles) formed without the participation of macromolecules and free of DDS molecules.

COPOLYMERS OF N-VINYLPYRROLIDONE CONTAINING IONIC GROUPS (BY LUMINESCENCE METHOD)

The methods used for studying the complexation reaction between PVP with ionic surface-active substances (SAS) (conductometry, viscosimetry, etc.) do not give complete information on interactions of macromolecules with SAS molecules because of the high sensitivity of this process of interaction to a large number of factors (concentration, temperature, etc.).

The method of luminescence was used by Anufrieva and co-workers [235–237] for the study of this interaction with an extensive set of VP copolymers containing functional groups of various structures since it allows investigation of the polymer complexation in diluted solutions.

The application of VP as a base monomer for preparation of such copolymers is caused by the following:

1. This monomer easily enters a copolymerization with a large number of monomers;
2. The copolymers obtained in a wide range of compositions are water-soluble;
3. The PVP chainlinks in the copolymer do not hinder the study of electrostatic effects at the interaction of SAS and ionic group on the chains.

Structure of these copolymers with covalently connected fluorescent labels and of SAS are submitted as follows:

PVP-F-1

Copolymer of VP and methacrylic acid (CP–VP–MAA–F) containing 8–40 mol% of MAA.

$$\left[-CH_2-CH\right]_m^- \left[-CH_2-CH-\right]_n^- \left[CH_2-CH-\right]$$

with N–C=O (pyrrolidone ring), COOH, and C=O–O–CH$_2$–(anthracene) substituents

Copolymer of VP and allylamine (CP—VP—AA—F) containing 7—31 mol% of AA.

$$\left[-CH_2-CH\right]_m^- \left[-CH_2-CH-\right]_n^- \left[CH_2-CH-\right]$$

with N–C=O (pyrrolidone ring), CH$_3$/COOH, and CH$_3$/C=O–O–CH$_2$–(anthracene) substituents

—(CH$_2$—AnT)

Copolymer of VP and crotonic acid (CP–VP–CA–F) containing 12–50 mol% of CA.

$$\left[-CH_2-CH\right]_m^- \left[-CH_2-CH-\right]_n^- \left[CH_2-CH-\right]$$

with N–C=O (pyrrolidone ring), NH$_2$, and NH—C(=O)—NH—CH$_2$—AnT substituents

Copolymer of VP and vinylamine (CP–VP–VA–F) containing 9–13 mol%

$$\left[-CH_2-CH\right]_m^- \left[-CH_2-CH-\right]_n^- \left[CH_2-CH-\right]$$

with N–C=O (pyrrolidone ring), CH$_2$–NH$_2$, and CH$_2$—NH—C(=O)—NH—CH$_2$—AnT substituents

Copolymer of VP and allylamine (CP–VP–AA–F) containing 8–18.5 mol%

$$\left[-CH_2-CH\right]_m^- \left[-CH_2-C\right]^- \left[CH_2-CH-\right]$$

with N–C=O (pyrrolidone ring); CH$_3$/C=O–O–(CH$_2$)$_2$—N(CH$_3$)$_2$; and CH$_2$—NH—C(=O)—NH—CH$_2$—AnT substituents

Copolymer of VP and *N,N*-dimethylaminoethylmethacrylate (CP–VP–DMEMA–F), containing 12.5 mol%

$$\left[-CH_2-CH\right]-\left[-CH_2-\underset{\underset{O}{|}}{\overset{\overset{CH_3}{|}}{C}}-\right]-\left[CH_2-CH-\right]$$

with pyrrolidinone ring $N{-}C{=}O$; methacrylate side: $\overset{CH_3}{\underset{|}{C}}{=}O$, O, CH_2, $CH_3{-}N^+{-}CH_3$, R_1; and $CH_2{-}NH$, $C{-}NH{-}CH_2{-}AnT$, O

$R = C_nH_{2n+1},$ $n = 1, 2, 4, 10, 12, 14$

Copolymer of VP and N,N-dimethyl-N-alkylammoniumethylmethacrylate (CP–VP–DMAEM–F) containing 12 mol%

$$\left[-CH_2-CH\right]-\left[-CH_2-\underset{\underset{O{-}(CH_2)_2{-}N^+}{|}}{\overset{\overset{CH_3}{|}}{C}}-\right]-\left[CH_2-CH-\right]$$

with $CH_2{-}NH{-}C{-}NH{-}CH_2{-}AnT$, O; and C_2H_5, Re, C_2H_5 on N^+

$Re = C_nH_{2n+1}$ $n = 1, 2, 4, 10, 12$

Copolymer of VP and N,N-diethyl-N-alkylammonium ethylmethacrylate chloride (CP–VP–DEAEM) containing 11 mol%.
Dimethyl benzyl alkylammonium chloride (DMBAA):

$$\langle\text{phenyl}\rangle{-}CH_2{-}\underset{\underset{CH_3}{|}}{\overset{\overset{CH_3}{|}}{N}}{}^+{-}C_nH_{2n+1}$$

where $n = 10, 12, 14$.
Sodium alkylsulfate (AS):

$$C_nH_{2n}{-}O{-}SO_3^+Na^-$$

where $n = 10, 12, 14$.
And sodium salt diisoocthyl sulfosuccinate structure (DOS):

$$C_4H_9{-}\underset{\underset{?}{}}{\overset{\overset{C_2H_5}{|}}{CH}}{-}CH_2{-}O{-}\underset{}{\overset{\overset{O}{\|}}{C}}{-}CH_2$$
$$C_4H_9{-}\underset{\underset{C_2H_5}{|}}{CH}{-}CH_2{-}O{-}\underset{\underset{O}{\|}}{C}{-}CH{-}SO_3^-Na^+$$

It was found [235–237] that the relaxation times τ_w, describing intramolecular mobility of macromolecules in a solution at given temperature T and solvent viscosity (equation 4.6), and polarization degree of luminescence of the fluorophor on a chain (equation 4.7) prove to be responsive to the interaction of the macromolecule with SAS ions

$$\tau_w = (1/P' + 1/3)\, 3\tau_0/(1/P + 1/P') \qquad \text{(eq. 4.6)}$$

where P is the value of luminescence polarization of the labelled polymer τ_0 the duration of luminescence and $1/P$ a parameter corresponding to the amplitude of high-frequency movements of the fluorophor or thermodynamic flexibility of a chain connecting an anthracene ring with the main polymer chain.

$$P' = (J_{\mathrm{II}} - J_\perp)/(J_{\mathrm{II}} + J_\perp) \qquad \text{(eq. 4.7)}$$

where J_{II} and $J_\perp$ are luminescence intensities of the light vector parallel (J_{II}) and perpendicular ($J_\perp$) to the light vector.

As a parameter reflecting change of intramolecular mobility at complexation reaction it is proposed to use the reciprocal value of polarization $(1/P)$ luminescence of an anthracene-labelled polymer [235]. It is also stated that the identical degree of filling in sites of macromolecules by SAS ions at various concentrations of ionic groups on a chain (C_{ION}) brings about identical changes of intramolecular mobility in the polymer molecule.

It was developed the determination method of dissociation constants $(K_D = 1/K_b)$ or binding (or stability) constants (K_b) polymer + SAS complexes and the degree of filling of macromolecular chains as carriers of SAS ions in a solution by a method of polarized luminescence (PL) [235].

The dissociation constant (K_D) of the complex of a SAS molecule with an ionic group of a VP copolymer is described by the relation

$$K_D = C'_{\mathrm{ION}}\, C'_{\mathrm{SAS}}/C_{\mathrm{ION-SAS}} \qquad \text{(eq. 4.8)}$$

where C'_{ION} and C'_{SAS} are concentrations of ionic groups of copolymer and SAS respectively, not included in the complex, $C_{\mathrm{ION-SAS}}$ is a complex concentration in a solution.

The K_D value is determined by the following expression:

$$K_D = C_{\mathrm{ION}} \cdot (\beta - \theta) \cdot (1 - \theta)/\theta \qquad \text{(eq. 4.9)}$$

where C_{ION} and C_{SAS} are common concentrations of ionic groups of copolymer and SAS in a solution, $\beta = C_{\mathrm{SAS}}/C_{\mathrm{ION}}$ is a ratio of concentrations of SAS ions and of ionic polymer groups, $\theta = C_{\mathrm{ION-SAS}}/C_{\mathrm{ION}}$ is a fraction of ionic groups of copolymer filled by SAS ions, $C'_{\mathrm{ION}} = C_{\mathrm{ION}} - C_{\mathrm{ION-SAS}}$ and $C'_{\mathrm{SAS}} = C_{\mathrm{SAS}} - C_{\mathrm{ION-SAS}}$.

The relation (eq. 4.10) was obtained by determining θ using eq. 4.9 for the same system of copolymer–SAS at the various contents of polymer and ionic groups following from equality of the K_D values at identical filling degrees of

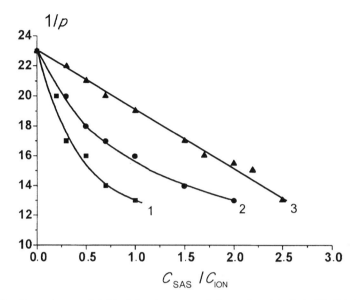

Fig. 4.7. Dependence of $1/P$ of luminescene of fluorophor labelled CP–VP–CA (19.4 mol%) copolymer on SAS (DMBA) concentration: 33×10^{-4} (1), 7×10^{-4} (2), and 3.2×10^{-4} (3), pH 9.0, 25 °C [235]. Reproduced by permission of Nauka.

ionic groups in copolymer by SAS molecules [230]:

$$\beta_1 = (C_{ION}^1 \beta_1 - C_{ION}^2 \beta_2)/(C_{ION}^1 - C_{ION}^2) \qquad \text{(eq. 4.10)}$$

where C_{ION} and β with indexes 1 and 2 are the characteristics of solutions of copolymer–SAS systems, where the value of θ proves to be identical.

Thus, using external parameters of the system, namely C_{ION} and β, it is possible to estimate internal parameters (K_D and θ) characterizing the ability of the components to complex formation (or dissociation) and complex composition.

Determination of the $1/P$ values for solutions with various β, achievable the concentration variation of polymer and SAS, allows the estimation the values of C_{ION}^1 and C_{ION}^2 adequate to identical values of θ or $1/P$ (Fig. 4.7).

From relations (eq. 4.9 and eq. 4.10) the value of a filling degree of macromolecules and dissociation constants (K_D) of copolymer–SAS complexes was calculated. The typical dependence of K_D on θ in half-logarithmic scale (CP–VP–CA–F + DMBA in water) is shown in Fig. 4.8.

Pautov and co-workers established [238] that in a wide range of θ (0.2–0.8) the dependence $\log K_D$ on θ for the majority of polymers investigated is linear. Deviations of the dependence of $\log K_D$ on θ in a range of small and rather large concentrations of SAS is explained positions of the complex equilibrium theory [239], when K_D becomes dependent on the filling degree:

$$K_D = K_D^{ch} \, e^{-\varphi} \, (\theta) \qquad \text{(eq. 4.11)}$$

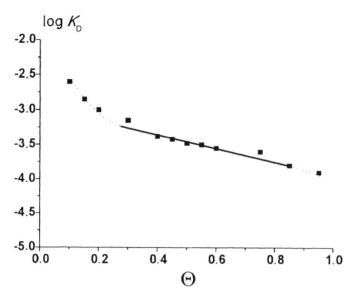

Fig. 4.8. Dependence of K_D on θ for complexes of copolymer CP–VP–CA (19.4 mol%) and DMBA in water, pH 9.0, 25 °C [238]. Reproduced by permission of Nauka.

where K_D^{ch} is a characteristic dissociation constant and, φ (θ) an arbitrary function. If φ (θ) is a decreasing function, the SAS ion binding by one site of the polymer-carrier hampers that by others. If φ (θ) is a growing function the binding of SAS ions by separate sites of macromolecule results in a strengthening of the SAS binding by other sites.

In relation to

$$\log K_D = \log K_D^{ch} - \varphi\,(\theta) \log e \qquad \text{(eq. 4.12)}$$

the φ (θ) function determines the character of dependence of $\log K_D$ on (θ). Two parameters are used, namely $\log K_D^{ch}$ and numerical coefficient (φ) at θ under sign of exponent in eq. 4.11 for the complete description of K_D in a range from $0.2 < \theta < 0.8$ [236].

The φ coefficient is characteristic of a binding co-operativity of SAS ions by the copolymer macromolecules. The K_D^{ch} is received by an extrapolation to zero for a linear section of the dependence of $\log K_D$ on θ in a range $0.2 < \theta < 0.8$.

The physical sense of K_D^{ch} is that the factor reflects affinity for a separate SAS ion to an ionic group of the copolymer in terms of the contribution of hydrophobic interactions of SAS alkyl radicals contacted by a macromolecule. The value of K_D^{ch} received by extrapolation to zero value of θ of the section of curve between $0 < \theta < 0.2$ indicates mainly electrostatic interactions.

Table 4.6   The equilibrium dissociation constants (K_D) of complexes of copolymers with different ionic group contents (C_{ION}) and SAS in water at $\theta = 0.2$ and 0.8, φ and K_D^{ch} at 25 °C [236]

No.	Copolymer+SAS system	C_{ION} (mol%)	SAS chain length	φ	$K_D^{ch} \times 10^4$ (mol/l)	$1/K_D^{ch} = K_b$ (l/mol)	$K_D \times 10^4$(mol/l) $\theta = 0.2$	$\theta = 0.8$
1.	CP-VP-CA+DMBAA	13.3	12	2.9	65	155	37	6.6
2.		15	12	2.6	29	350	17	3.5
3.		19.4	10	1.5	24	420	24	22
4.		19.4	12	2.4	14	720	8.8	2.1
5.		19.4	14	1.2	0.89	11 200	0.7	0.34
6.		30.8	12	2.3	12	830	7.4	1.8
7.		40	12	2.9	29	530	11	1.9
8.		50	12	3.1	29	345	15	2.5
9.	CP-VP-AA+DMBAA	7	10	1.5	28	355	21	8.2
10.		7	12	0.46	10	1 000	9.1	6.9
11.		17	10	2.2	48	210	31	8.1
12.		17	12	0.74	12	830	10	6.6
13.		31	10	3.2	52	190	27	4
14.		31	12	5.2	110	90	39	1.7
15.	CP-VP-MAA+DMBAA	9.7	12	1.6	4.8	2100	3.5	1.4
16.		14.5	12	2.8	7.1	1 400	4	0.72
17.		40	12	4.4	24	2 280	9.9	0.70
18.	CP-VP-VA+AS	9	10	3.0	27	370	15	2.4
19.		9	12	1.7	2.3	4300	1.6	059
20.		9	14	4.7	0.63	16000	0.25	0.015
21.		12.8	12	2.2	4.3	2300	2.8	0.74
22.	CP-VP-AA+AS	8	12	3.1	59	170	32	5.1
23.		18.4	12	0.75	3.21	3 100	2.8	1.8
24.	CP-VP-AA+AS	18.4	14	1.0	0.65	15500	0.53	0.28
25.	CP-VP-DMEMA+AS	12.5	12	2.1	11	910	7.2	2.1
26.	CP-VP-DMEMA-Re(1)+AS	12	12	3.5	180	550	89	11
27.	CP-VP-DMEMA-Re(14)+AS	12	12	4.9	5	2000	2	0.1
28.	CP-VP-DEAEM+AS	11	12	1.9	14	720	10	3.1
29.	CP-VP-DEAEM-Re(2)+AS	11	12	1.3	49	200	38	17
30.	CP-VP-DEAEM-Re(14)+AS	11	12	2.3	59	170	37	9.4

The parameters of complexation for a larger number of copolymers submitted above with SAS are given in Table 4.6. On the basis of the analysis of these data in comparison to results of the study of intramolecular mobility of polymer carriers for SAS ions [236] the factors influencing stability, composition and structure of copolymer–SAS complexes are revealed.

These factors are distribution of charged groups on chain of the polymer–carrier and the length of alkyl chains of SAS ions. Therefore, the stronger binding of SAS molecules takes place for copolymers with a microblock distribution of charged groups on chain in comparison with statistical one (see Nos. 2, 16, 19 and 21 in Table 4.6). The high stability constants are characteristic of VP copolymers with methacrylic acid or N-vinylamine where distribution of chainlinks is microblock. Copolymers of VP + crotonic acid (CA) or of VP + allylamine (AA) having statistical distribution of chainlinks possess lower values of stability constant (K_b).

The higher ability to complex formation of copolymers with microblock distribution of charged groups is connected so that a more complete realization of hydrophobic interactions is sterically possible between the bound alkyl SAS chain being near to the copolymer chain.

Schematically, the elements of structural complexes: (a) microblock distribution of groups; (b) statistical distribution can be as follows:

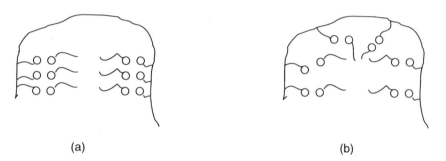

(a)   (b)

At increase of number of carbon atoms from 10 to 14 in linear SAS chains bearing both positive and negative charge, stability constants of the polymer-SAS complexes greatly increases (compare K_b for Nos. 3, 4 and 5 in Table 4.6) from 420 (C_{10}) to 11200 l/mol (C_{14}). Such increase of K_b is defined by strengthening of hydrophobic contacts between alkyl SAS chains due to lengthening of SAS alkyl fragments.

A role of branching in SAS alkyl chains on complex stabilization was displayed. Therefore, CP–VP–AA (18.4 mol%) forms the more stable complex with DOS (SAS with branched structure) then dodecylsulfate (DDS) molecules (Nos. 23 and 24).

The stability of complexes between copolymers and SAS is defined by the contents of ionic groups in the copolymer. For complexes of SAS (DMBAA) ions with static copolymer of CP–VP–CA–F, the dependence of K_D and K_D^{ch} on the composition of copolymer is shown, having $K_{D\,min}$ at 30 mol% of CA in the copolymer (Nos. 1, 2, 4, 6–8 in Table 4.6 and Fig. 4.9).

Such behaviour of K_D at change of copolymer composition is caused by the action of opposite working forces. Therefore, probability of contact formation at the expense of hydrophobic interactions between SAS ions connected to the polymer-carrier grows only so long as the electrostatic repulsion of charged groups on a chain does not begin to prevent the approach of fragments separated on chain [236–238].

The effect of side substituent structure of VP copolymers bearing positively charged groups on stability of complexes can be analysed in CP–VP–VA,

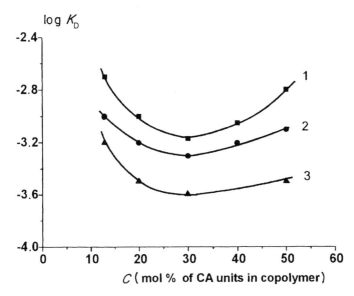

Fig. 4.9. Dependence of K_D of copolymer of CP–VP–CA + DMBAA complexes on the copolymer composition at $\theta = 0.2$ (1), 0.4 (2) and 0.7 (3) in water, pH 9.0, 25 °C [236]. Reproduced by permission of Nauka.

CP–VP–DMAEM, CP–VP–DEAEM, CP–VP–DMAEM–R and CP–VP–DEAEM–R, interacting with DDS ions. The replacement of side substituents increasing their volume and removing charges from the main chain brings about destabilization of the polymer + SAS complex (Nos. 19, 25 and 29 in Table 4.6).

Introduction of an alkyl group with 14 carbon atoms in CP–VP–DMAEM–R causes some stabilization of complexes for this copolymer +AS (Nos. 25–27 in Table 4.6), whereas the stability of CP–VP–DEAEM–R + AS complexes thus does not, in practice, change or even slightly decrease (Nos. 25–30 in Table 4.6) possibly due to a steric screening of charged groups in the polymer chains at the complex formation.

IONIC STRENGTH EFFECT ON SAS–COPOLYMER COMPLEXATION

The effect of the introduction of salt (NaCl) on the stability of the complexes is defined by peculiarities of copolymer composition co-operating with SAS ions [235]. It is that the screening of charges by salt ions, on the one hand, favours weakening of an electrostatic attraction between opposite charges of groups in the polymer chain and groups in the SAS molecule (destabilization) and, on the other hand, decreases the forces of electrostatic repulsion between like charges on the chain (stabilization).

Table 4.7 The equilibrium dissociation constants (K_D), φ and K_D^{ch} of copolymer + SAS complexes in water–salt solutions [236]. Reproduced by permission of Nauka

System type	Copolymer + SAS system	Ionic strength	φ	$K_D^{ch} \times 10^4$ (mol/l)	$K_D \times 10^4$ (mol/l) $\theta = 0.2$	$\theta = 0.8$
I	CP–VP–CA	Water	2.4	14	8.7	2.1
	(19.4 mol%)	0.1 N NaCl	4.4	110	46	3.3
	DMBAA ($n = 12$)	0.2 N NaCl	6.3	120	34	0.8
II	CP–VP–VA					
	(13 mol%)	Water	2.2	4.3	2.8	0.74
	+ DDS($n = 12$)	0.1 N NaCl	3.4	26	13	1.7
		0.2 N NaCl	3.5	25	11	1.1
		0.3 N NaCl	3.4	17	86	1.1
		0.1 N MgCl$_2$	2.8	9	5.1	0.8
		0.2 N MgCl$_2$	4.2	1.8	7.8	0.6
III	CP–VP– DMEAM–Re(1)	Water	3.5	180	89	11
	(13 mol%) + DDS	0.1 NaCl	3.9	68	31	3
		0.2 NaCl	40	58	26	2.4
IV	CP–VP–DEAEM	Water	1.3	49	38	17
	– Re(2) (11 mol%)	0.1NaCl	4.0	77	35	3.1
	+ DDS	0.2 N NaCl	4.1	72	32	2.7
		0.3 N NaCl	3.9	49	22	2.2
V	CP–VP–	Water	2.3	59	37	9.4
	DEAEM–Re(1)	0.1 N NaCl	4.1	73	32	2.7
	(11 mol.%)	0.2 N NaCl	4.1	45	20	1.7
	+ DDS ($n = 12$)					

The set of systems where the first factor dominates involves I and II systems in Table 4.7. In these systems the addition of salt at small θ disturbs the complex.

For systems in which the important role is played by hydrophobic interactions, the addition of salt strengthens the interaction stabilizing the complexes through the second of the specified mechanisms (III and IV systems in Table 4.7). In these systems the binding co-operativity (φ) of SAS ions by polymer-carrier grows with the increase of ionic strength in the aqueous solution.

In the case of strong electrostatic interactions between oppositely charged groups being in chain and in SAS ions, the contribution of pyrrolidone rings or hydrate shell of the chain in forming complexes can be insignificant. However, the manifestation of their role can be followed for the CP–VP–VA and CP–VP–AA copolymers at interaction with DDS (C_{12}), when K_b for copolymer with vinylamine (4300 l/mol) greatly exceeds that of copolymer with allylamine (170 l/mol) at the same composition of these copolymers (Nos. 19

and 22 in Table 4.6). In the former in which the $-NH_2$ group are near to chain, there are the probable additional interactions with participation of polarized water molecules forming hydrogen bonds with SO_3^- of DDS.

Thus, use of the fluorescent label attached to chain of VP copolymers has allowed Anufrieva and co-workers [230–232] to develop an original method for estimation of complexation constants for copolymers on the basis of VP and other monomers containing ionic groups with ionic SAS molecules in dilute aqueous solutions, and to gain insight into the factors promoting stabilization or destabilization of such types of complexes.

4.5 SYNTHETIC POLYMERS

SULFONATE-CONTAINING AROMATIC POLYAMIDES

The formation of water-soluble polymer–polymer complexes was found by a method of polarized luminescence [240] on an example of PVP (or anthracene labelled PVP, PVP–F) and aromatic polyamide with two sulfonic groups (PA-2) (or anthracene labelled polyamide, PA-2-F). The structure is as follows:

PA-2, sodium salt of poly-4,4'-diphenyl-iso-phthalamide-2.2'-disulfonic acid.

Mobility of a terminal fragment in PA-2-F in a dilute aqueous solution (0.02–0.04 wt%) is characterized by the value of relaxation time (τ_w) being 4 ns. On the addition of salt to the solution (0.5 N NaCl) the intramolecular mobility of a macromolecule ΠA-2-F becomes more restrained, with the value of τ_w increasing from 4 to 7 ns in water and to 12 ns in acidic (pH = 2.5) solution.

As pK_a for sulfonic acid groups in PA-2-F is 2.9 [241], a large part of the groups will be in H^+ form. On addition of PVP to a aqueous solution containing PA-2-F (pH = 2.5) τ_w of a polyamide molecule increases to 17 ns. At the same time the τ_w of PVP macromolecules in this system with PA-2 also rises more significantly from 27 to 140 ns. These findings confirm the formation in a solution of an interpolymer complex between macromolecules of polyamide and PVP.

Introduction of NaCl in solution in which the macromolecules of two polymers are present increases the τ_w for PA-2-F from 17 to 170 ns, indicating a tightly polymer complex formation. Simultaneously the relaxation times describing intramolecular mobility of PVP (anthracene-labelled PVP), decreases up to the τ_w values characteristic of free PVP chains in a solution (27 ns).

The set of these data testifies that at shielding of charges (ionized sulfonic groups) on a polyamide chain in competitive intermolecular interactions between both polyamide–polyamide and polyamide–PVP the former begins to prevail with the formation of $(PA-2)_n$ permolecular aggregations and with PVP releasing in a solution.

Thus, in the system investigated PVP plays the role of gel-forming initiator, promoting the occurrence of permolecular $(PA-2)_n$ aggregations from aromatic polyamide macromolecules even at $C = 0.04$ mol% in solution without entering the oligomer associate thus formed.

The formation of permolecular $(Pa-2)_n$ aggregations on the PVP macromolecule as a matrix in a water–salt solution and the subsequent transition of these from the complex in a solution find a reflection not only in the increase of the τ_w in the case of PA-2 from 17 to 170 ns and a reduction in the τ_w values in the case of PVP to that being characteristic of free PVP chains but also in a number of optical and spectral changes for each of the components of this system [240].

Thus, in this example of PA-2 + PVP an interesting phenomenon was displayed concerning the aggregation of polyamide structure macromolecules in the presence of poly-*N*-vinylamide macromolecules in dilute aqueous solutions. Sulfonate-containing macromolecules could approach closely enough to PVP macromolecules for interaction to be sufficiently strong between them to be responsible for forming a $(PA-2)_n$ aggregate of a final structure without a PVP initiator.

The intermolecular complex from macromolecules of a partially ionized PA-2 polyamide containing $—SO_3^-$ and $—SO_3H$ groups and PVP in water arise due to the interaction between polarized molecules of water in a hydrate shell of PVP through hydrogen bonds of type $—SO_3H\cdots O—H\cdots O—H\cdots O{=}C—$ and $SO_3^-\cdots H—O\cdots H—O—H\cdots O{=}C—$. Thus near to PVP macromolecules with a high average degree of polymerization ($P = 300$) accumulation of short chains of PA-2 polyamide macromolecules ($P = 15–20$) occurs to promote strengthening of the interaction between them with participation of $C{=}O\cdots H—N—$ groups of various chains.

The introduction of chloride (NaCl) ions in this system increases the concentration of those near to the PVP chain because of the formation of weak complexes (see section 4.1) with participation of water molecules of hydrate shell and, as a result, causes the replacement of PVP macromolecules in solution from the PA-2 + PVP complex. The same effect can also be due to the presence of Na^+ cations, since these decrease the concentration of ionized sulfonate groups participating in a network of hydrogen bonds near to the PVP chain. In the presence of chloride ions, the aggregates from polyamide macromolecules bearing a large number of ionic groups generated under the influence of PVP macromolecules are separated from the latter, made secure at the expense of an increase of ionic strength and pass in a solution without decomposing.

POLYCARBOXYLIC ACIDS

The interaction of macromolecules of a polyacrylic (PAA) or polymetacrylic (PMAA) acids and PVP in a aqueous solution resulting in the formation of a polymer–polymer complex is claimed [242, 243] to be due to hydrogen bonds between C=O of PVP and carboxylic groups in the protonic form.

In dilute aqueous solutions (5 wt%) at certain pH value, when the greater part of the carboxylic groups of polyacid is in a unionized state, these two polymers (1:1 in mole) co-operate and precipitate in a separate phase from the solution. At the same mole ratio of PVP and low-molecular-weight analogue of a chainlink of polyacid (acetic, butyric or dimethylacetic acids can act as analogue of these acids) in a aqueous solution the precipitation of polymer does not occur, though the weak complexation of PVP macromolecules with these small molecules takes place in water (see Chapter 3).

The complexation of two macromolecules occurs through a few stages. At the first stage an intersegment movement of chainlinks on flexible-chain macromole-cules with the diffusion constant ($K_{dif} = 10^7–10^8$ 1/mol s) which is 10–10^2 times lesser than that for low-molecular-weight analogues of the chainlinks ($10^9–10^{10}$ 1/mol s [244]) promotes the approach of several chainlinks on length of a segment from unlike macromolecules with simultaneous partial replacement of water molecules from hydrate shells of those forming a two-cord structure. In the second stage the 'two-cord' structures are formed in large numbers, being accompanied by shrinkage of co-operating macromolecules and the squeezing out of hydrate water. As a result, in the third stage, the dense aggregates (similar to tertiary structure in proteins) in which enter many fragments into the 'two-cord' structure [245] are formed.

Clearly, such a shrinkage of interacting macromolecules finds a reflection in the segmental mobility of chain fragments included in the polymer–polymer complex.

A change of intramolecular mobility in chains of water-soluble polymers occurs at such an interaction and is followed by a polarized luminescence method with application of anthracene-labelled polymers. As a polymer object it was in used anthracene-labelled polyvinylamides (PVP–F, PVCL–F and PVMA–F) and polyacids such as PAA (PAA–F) and PMAA (PMAA–F) [246, 247].

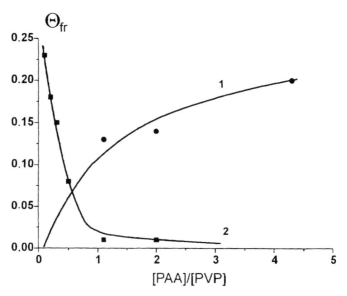

Fig. 4.10. Dependence of the fraction ($\theta_{free} = 1 - \theta_{ppc}$) for polymer fragments of PVP (1) and PAA (2) uncombined in the complex in water on the mole ratio of components used for formation of the complex, [PVP] = 0.1% [241]. Reproduced by permission of Nauka.

The fraction of chains or chain fragments (θ_{ppc}) in each of polymers included in a polymer–polymer complex (PPC) at different mole ratio (γ) of co-operating polymers is determined by the following equation:

$$1/\tau = \theta_{ppc}/\tau_{ppc} + (1 - \theta_{ppc})/\tau_0 \qquad \text{(eq. 4.13)}$$

where τ is the measurable relaxation time characterizing the intramolecular mobility of chains of a component shown by the label in solution and τ_0 and τ_{ppc} the relaxation times describing those of free polymer chains in a solution and chains of the same polymer included in a polymer–polymer complex.

The composition of this complex is determined by the value of γ at which the fraction of chain fragments not combined in the polymer–polymer complex for each of the polymers was minimal (Fig. 4.10) [246].

The actions of polymer nature and solvent nature on the relaxation times of complexes and their composition are represented in Table 4.8.

All poly-*N*-vinylamides investigated interact with non-ionized polycarboxylic acids (PAA or PMAA) and form polymer–polymer complexes in various solvents (Table 4.8); there is substantial growth of relaxation times of chain fragments testifying to their interactions. The complex composition in water is equal to 2:1, meaning that one chainlink of the polyacid accounts for two poly-*N*-vinylamide chains. At the same time the mole ratio between chain-

Table 4.8 The relaxation times describing intramolecular mobility polymer chains included in a complex (τ_{ppc}^{sp}) and free (τ_0^{sp}) in various solvents at optimum composition (γ) of the complex (25 °C) [246]

	τ_{ppc}^{sp} (ns)	τ_0^{sp} (ns)	τ_{ppc}^{sp} (ns)	τ_0^{sp} (ns)	τ_{ppc}^{sp} (ns)	τ_0^{sp} (ns)	τ_{ppc}^{sp} (ns)	τ_0^{sp} (ns)
	Water ($\gamma = 0.5$)		Methanol ($\gamma = 1.0$)		DMF ($\gamma = 1.0$)		N-methyl-pyrrolidone ($\gamma = 1.0$)	
Composition								
PMAA–F + PVA	430	30	—	5.5	30	5.3	5.6	5.6
PMAA + PVP–F	360	11	270	3.3	20	2.4	2.5	2.5
PAA + PVP–F				2.9	18	3.5	3.5	3.5
PAA–F + PVP	120	11	420	3.3	10	2.4	2.5	2.5
PAA–F + PVCL		10	—	2.9	3.6	3.5	3.5	3.5
PAA + PVCL–F	120	20	150	6.3	5.9	5.5	5.5	5.5
PMAA–F + PVCL	430	30	—	5.5	40	5.3	5.6	5.6
PMMA + PVCL–F	120	20	160	6.3	26	5.5	5.5	5.5
PMAA–F + PVMA	130	30	—	—	—	—	—	—
PMAA + PVMA–F	130	10	—	—	—	—	—	—

Notes: The values of $\tau^{sp} = \tau (\eta^{sp}/\eta)$ are reduced to solvent viscosity ($\eta^{sp} = 0.38$ sP), $\gamma = $ [polyacid] : [poly-*N*-vinylamide], [polymer] = 0.05–0.5%.

links of various macromolecules in the complex formed in methanol or DMF is dissimilar and corresponds to 1:1.

Further, the mobility of chains in the complex of poly-*N*-vinylamide and polycarboxylic acid in water is strongly restrained: the value of τ_{ppc} in comparison with τ_0 increases by 10 and more times (see Table 4.8). In this case of chain fragments labelled by anthracene ring in both PMAA and PVP is immobilized in dense formations made of a greater number of fragments of various macromolecules.

It should be noted the considerable distinction of the τ_{ppc} value for various polymer pairs (130–430 ns) indicating, probably, the indirect manner of the various contents of residual hydrate water in complexes. Accordingly, for a pair of PVP + PAA ($\tau_{ppc} = 120$ ns) the contents of water exceed those for a pair of PVCL + PMAA ($\tau_{ppc} = 430$ ns).

It follows from data in Table 4.8 that in methanol poly-*N*-vinylamides interact with polycarboxylic acids and form polymer–polymer complexes. The mobility of chains in such complexes is much restrained: the relaxation times (τ_{ppc}) increase from $\tau_0 = 3$–5 ns to $\tau_{ppc} = 200$–400 ns. It is of interest that the introduction of methanol to aqueous solutions does not change the chain fragment fraction (θ_{ppc}) included in a complex: the complex retains in mixed water–methanol solutions (Fig. 4.11). However, a compact (tertiary) structure generated in water begins to be destroyed with increasing methanol concentration in a mixture (Fig. 4.11) that is supported by an increase of mobility of the co-operating chains. The τ_{ppc} value falls from 800 ns for pure water to 190 ns for 20 mol% of methanol in water and to 280 ns in a pure methanol. It suggests [246] that in methanol the complex holds a two-cord structure.

All amide solvents investigated (DMF and N-methylpyrrolidone) destroy not only the tertiary structure of polymer–polymer complexes (Figs. 4.11, 4.12 and 4.13), but also the secondary structure, since they partially or completely prevent the formation of intermolecular contacts.

In an aqueous solution N-methylpyrrolidone destroys these complexes to a greater degree than DMF or methanol (Fig. 4.11). In fact, the value of τ_{ppc} for a pair from PVP–F and PMAA decreases about 10 times after introducing as little as 5–6 mol% of the solvent in water, whereas it reduces four times when the methanol concentration achieves 20 mol%. In addition, the tertiary dense structure of the PVP + PMAA complex disappears on the introduction of the above mentioned N-methylpyrrolidone concentration (Fig. 4.11). However, the secondary (two-cord) structure of the same complex becomes disrupted and is complete only at 50 mol% of this solvent in water (Fig. 4.12).

As seen in Figs. 4.11 and 4.12 methanol, unlike N-methylpyrrolidone, acts on the tertiary structure only after reaching 20 mol% with retaining the two-cord structures at more high methanol concentrations (60–70 mol%).

The role of poly-N-vinylamide structure on examples of PVP and PVCL and acids (PAA and PMAA) on complex stability to action of organic solvent is well illustrated in Fig. 4.13. In the system of PMAA + PVCL the complexes retain up

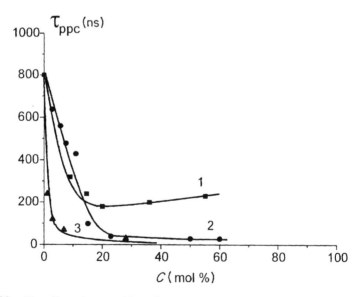

Fig. 4.11. The effect of composition of water solvent mixture (methanol-1, DMF-2 and N-methylpyrrolidone-3 in mol%) on intramolecular mobility (τ_{ppc}) of PVP + PMAA complex [241]. Reproduced by permission of Nauka.

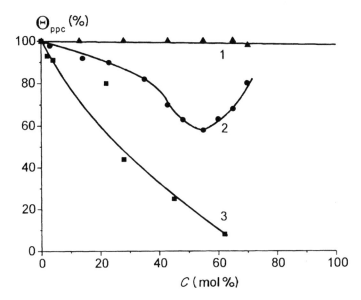

Fig. 4.12. Dependence of fraction (Θ_{ppc}) of chain fragments included in complex (PVP + PMAA) on composition of a mixture from water + organic solvent (methanol-, DMF-2 and *N*-methylpyrrolidone-3 in mol%) [241]. Reproduced by permission of Nauka.

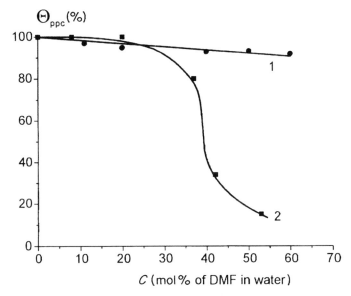

Fig. 4.13. Dependence of fraction of chain fragments included in the complex (Θ_{ppc}) on composition of water + dimethylformamide (mol%) mixture for PVCL-F + PMAA (1) and PVCL–F + PAA (2) [241].

to 60 mol% of DMF unlike those of PAA + PVCL which completely disappear already about 40 mol% (Fig. 4.13). DMF has a pronounced effect on the PMAA + PVP complex in comparison to the PMAA + PVCL system.

Thus, the study of fragment chain mobility in polymer–polymer complexes of poly-*N*-vinylamides and polyacrylic acids in solvents of various nature reveals a number of conformational structures of co-operating macromolecules. In an aqueous solution the interaction between macromolecules probably occurs in a few stages. At the first stage two-cord structures (secondary structure) between separate fragments of co-operating chains with partial replacement of hydrate water are built. At the subsequent stages these structures approach each other, resulting in shrinkage and hydrophobization with the formation of tertiary structure [247].

In such a solvent as DMF the molecules of which are capable of the formation of hydrogen bonds with polyacids in H⁺-form there are only the 'two-cord' structures. The formation of a 'two-cord' structure between PVP and PMAA in methanol (Fig. 4.12) is probably connected to the presence of extrinsic amounts of water in alcohol and in polymers.

THE EFFECT OF IONIZATION POLYACID DEGREE ON THE COMPLEX FORMATION

The destruction of a polymer–polymer complex can be achieved by increasing the number of ionized groups ($—COO^-$) on polyacid chains. Thus, as is well known [248], ionization of both PAA and PMAA releases significant forces of electrostatic repulsion along the chain with the unfolding of coils and the increase of intramolecular mobility of chainlinks. So, the τ_w^{sp} value decreases when the ionization degree of polyacid increases. It is particularly marked in PMAA unlike PAA in going from a protonated form ($\tau_w^{sp} = 33$ ns (PMAA) and 10 ns (PAA)) to the ionized form ($\tau_w^{sp} = 7.0$ ns (PMAA) and 4.3 ns (PAA)) [248].

The degree of PMAA ionization greatly influences a fraction (θ_{ppc}) of chain fragments of a polyacid in a complex with PVCL, PVP and PVMA (Fig. 4.14) [247]. The increase of the number of negatively charged groups in PMAA destroys the complex in various degrees, depending on the poly-*N*-vinylamide structure. As seen on Fig. 4.14, the most stable complexes when all fragments of polyacid are included in interaction with other polymers are observed for PVCL. Only at $\alpha = 0.6$ parts of PVCL + PMAA complexes begins to be destroyed, whereas in the case of PVP + PMAA (or PVMA + PMAA) disintegration of the complexes occurs at $\alpha = 0.2$.

Hydrophobic interactions between methylene groups of PVCL rings with methyl groups of PMAA and, probably, the decreased permittivity near to chains promoting the formation of strong hydrogen bonds between polarized water molecules close to poly-*N*-vinylamide chains and carboxylic groups in

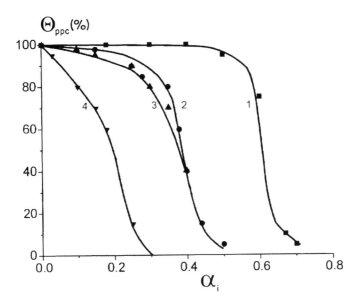

Fig. 4.14. The action of ionization degree of polyacid (α) on the fraction of chain fragments (Θ_{ppc}) in PMAA–F (1, 2, and 3) included in a complex with PVCL (1), PVP (2) and PVMA (3), and that of PAA–F (4), included in a complex with PVP (4) [247]. Reproduced by permission of Nauka.

H^+ form inhibit the disintegration of the 'two-cord' structures (Fig. 4.14). The substitution of PMAA by PAA (Fig. 4.14) in a complex with PVP greatly weakens the forces of interaction between macromolecules: for titration only 10 mol% of carboxylic groups are required, after which this complex begins to disappear.

Thus, poly-N-vinylamides such, as PVCL, PVP and PVMA in a aqueous solution with unionized PAA and PMAA form polymer–polymer complexes of various structural forms [249]. The most stable complexes with a complicated permolecular structure were established for a pair of PVCL and PMAA complexes.

In addition, their interaction is not limited by the formation of a two-cord structure (secondary structure) on various chain fragments occurring with the partial replacement of hydrate water from the shells of PMAA and PVCL. Further, these two-cord structures are thickened in more dense hydrophobic domains (tertiary structure). The latter is formed only in aqueous solutions.

A similar type of tertiary structure arises in the case of PMAA + PVP. However, it completely disappears at a small degree of ionization of carboxylic groups on a chain ($\alpha = 0.3$), while PMAA and PVCL tend to hold their structural formations at high ionization degrees up to $\alpha = 0.7$ [249].

Attention now focuses on the high stability of PMAA + PVCL complexes against PAA + PVCL, PAA + PVP or PMAA + PVP complexes. At the same time, if one can believe that between —COOH groups in PMAA and —C=O of amide groups in PVCL direct hydrogen bonds could be formed only in this system, the greatest steric problems would occur in the case of formation of these bonds because of the presence of voluminous substituents in both PVCL and PMAA. The most accessible for the formation of bonds of this type would be chains with small substituents, namely PAA and PVMA.

In a pair of PAA and PVMA these restrictions should be minimum in comparison with the pair considered above (PMAA + PVCL) that could result in formation of the stable complexes between them. However, it follows from Fig. 4.14 that this complex easily decomposes at small degrees of ionization of the polyacid.

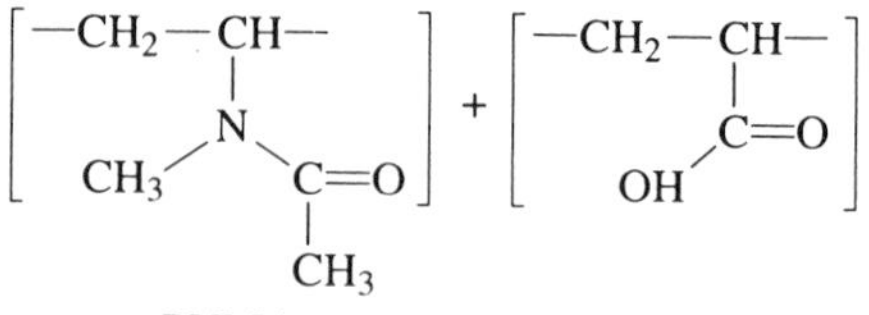

PVMA   PAA

Therefore these findings (see also Chapter 3) allow one to state that the complexation between macromolecules of poly-*N*-vinylamides and polyacrylic acids in an aqueous solution accurs due to the appearance of chain-like water associates connective COOH, on the one hand, and —C=O of amide groups, on the other, with the participation of C=O amide polarized water molecules. In this case the concept explains the reason of the absence of the proposed steric factor preventing direct hydrogen bond between the —COOH to the —C=O group when forming polymer–polymer complexes in the PMAA + PVCL system.

These binding water chains can be presented as follows:

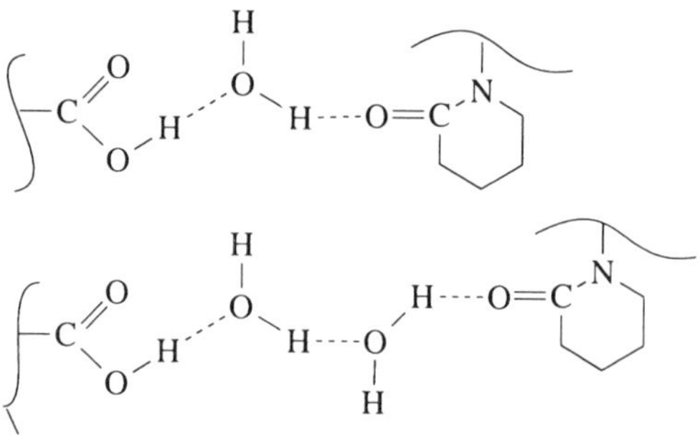

One can assume that chains of a similar type (hydrogen-associated chains with a large protic polarizability) were experimentally found in polyamide–phosphate systems (poly-L-histidine + KH_2PO_4[250] or poly-L-glutamic acid–NaH_2PO_4[251]).

These chains from molecules H_2O connective COOH groups of PMAA and C=O of PVCL are placed in poor polar locations with a low permittivity formed by caprolactam rings and methyl groups. The polymer–polymer complexes precipitated from water (poly-*N*-vinylamide + polycarboxylic acid) capture a certain amount of water (5–10 molecules per average number of chainlinks in a complex).

4.6 PROTEIN MOLECULES

One of the representatives of the poly-*N*-vinylamide generation, namely PVCL, possesses the ability to form complexes with protein molecules. This interaction is shown where the protein concentration in water, together with PVCL, falls sharply on solution heating to $T = 40\ °C$, when the isolation of the polymer occurs in a separate phase. On precipitating, the PVCL macromolecules capture protein molecules [252].

Introduction of PVCL with $M_w = 1.25 \times 10^6$ to solutions of proteins labelled with a fluorescent label (coproporphyrin) such as beef serum ovalbumin (BSO) with MW = 68 000) or rabbit immunoglobulin (IG with MW = 12 000) causes growth of quantum yield of a fluorescence of labelled proteins and the shift of the fluorescence maximum to 5 nm. Partial connectivity of the protein to polymer in a complex is established.

Heating an aqueous PVCL solution (0.5 wt%) and IG (10^{-8} mol/l) to 40 °C results in the formation of a polymer precipitate where the protein molecules are involved in a complex with PVCL. Thus, PVCL dissolves completely in the precipitate. If water is added to the filtered precipitate and the mixture allowed to cool to 20 °C, complete dissolution of polymer with the protein occurs. The findings [252] on the analysis of IG protein in the precipitate in the presence of various concentrations of NaCl and competitive BSO protein, show that in PVCL ($M_w = 1.25 \times 10^6$) up to 40% of the total amount of IG protein may be captured in the precipitate in the presence of NaCl 1.0 mol/l concentration. The introduction of BSO in the solution (3×10^{-6} mol/l) decreases a fraction of the absorbed IG protein in the precipitate to 14.0%, together with PVCL from a water (aqueous) solution.

The phenomenon observed of this PVCL co-precipitation of proteins or enzymes is of great interest in developing biotechnologocal processes as method of simple and convenient separation of protein systems from a reaction medium.

4.7 GELS OF POLY-*N*-VINYLCAPROLACTAM COMPLEXES

The ability of PVCL macromolecules with voluminous substituents to thermo-precipitate in a aqueous solution and to be isolated in a separate phase can be used as a basis for the design of the hydrogel forms as capsules or granules.

Therefore, the PVCL macromolecules posses this property in a range of temperatures of 30–34 °C, i.e. physiological temperatures (see Chapter 3) which cause no negative effects on the biological acitvity of proteins or the catalytic properties of enzymes. The other poly-*N*-vinylamides with long alkyl groups (propyl, butyl) have a low critical solution temperature in water at higher temperatures (48–60 °C).

Another important property of PVCL is that it interacts in aqueous solutions with an extensive set of compounds of various structures (low-molecular-weight and high-molecular-weight substances). Finally, an precipitation, PVCL macromolecules take biomolecules from an aqueous solution (and also other macromolecules), together with a large number of water molecules, creating conditions for the formation of hydrogel structures. In structures of such type, high mobility of small molecules and low mobility for macromolecules, for example, proteins, enzymes, cells, etc could be formed.

Combination of the properties specified shown above, PVCL or copolymers of VCL, has allowed new approaches to the formation of hydrogel granules from PVCL to be developed [253].

The formation of granules from PVCL is performed by addition of an aqueous solution of PVCL drop by drop in an aqueous solution containing a stabilizer at 40 °C. In a result, there is the formation of the granules of the spherical form.

The introduction of a PVCL solution in pure water provides formation unstable granules. These granules aggregate with stirring, and at decrease of temperature are completely dissolved. It is established that the value of the minimum concentration of PVCL in a solution at which granules are formed depends on polymer MW and decreases from 15 up to 4 wt% at the appropriate increasing MW in a range from 1.7×10^5 to 9×10^5.

Aggregation of granules and their dissolution has appeared possible in the reduction or elimination of substances at their introduction–stabilizers which are capable of complexation with PVCL macromolecules. As stabilizers it is possible to use phenols of various structures, polymers with carboxylic or sulfonic acid groups, SAS (dodecylsulfate), proteins (ovalbumin) and others.

Two methods of forming granules have been proposed (A and B). The first method (A) is that the stabilizer and PVCL are in one solution to which is then added, drop by drop, a heated solution of an addition agent (up to 40 °C water). In the second method (B) the PVCL solution involving the agent is added in a heated aqueous solution containing the stabilizer (40 °C). The second way (B) has appeared the most suitable, as it has permitted a choice of stabilizers promoting the formation of granules stable not only at 40 °C but also at room temperature. On contact of a drop (the PVCL solution) with a heated solution of the stabilizer, there occurs a simultaneous precipitation of macromolecules and their complexation in the superficial layer of a drop [253, 254].

Moreover, the PVCL precipitation on the surface of a drop causes strengthening of the complex formation between PVCL macromolecules and molecules of the stabilizer with origination of a dense superficial layer which fixes the spherical forms of granules and, probably, prevents the penetration of large molecules, both from inside the granules and from within the granules. A schematic structure of a granule, the size of which depends on a method of introduction (0.1 mm–1 mm in diameter), it is possible to present as a core of gel-like structure (nucleus) and a superficial dense layer from a complex PVCL + the stabilizer:

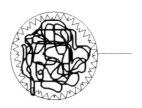

Adjusting the nature of the stabilizer or the concentration ratio of several stabilizers, granules stable at room temperature can be obtained and which are not sensitive to mechanical agitation.

In a solution of PVCL it is possible to enter a large group of low-molecular-weight compounds (amino acids, complexing agents and others), together with protein molecules (enzymes), live cells and microorganisms as the microcapsulation is formed in 'soft' conditions.

The top dense layer from PVCL molecules and the stabilizer itself represents a membrane between a nucleus of a granule and the external solution.

The effect of the stabilizer on the structure of this layer in the first 10 min of its formation is determined by the measurement of kinetic curves of a release from the granule of a low-molecular-weight compound, for example, sodium salt of glutamic acid (GA) (Fig. 4.15). As seen on Fig. 4.15, by use of phenols (tannin, resorcinol and phloroglucinol) the kinetic curves of the GA release at 40 °C are characterized by a sharp growth of GA concentration in an external solution during the first 10 min, with subsequent ceasing of its release from a granule.

Such behaviour is caused by the fact that the formation of a superficial layer (barrier layer for GA diffusion) from PVCL + phenols occurs during the same range of time from the beginning of the introduction of a drop in the heated solution.

For short time (10 min) the most dense layer in the presence of the phloroglucinol stabilizer is also formed: 60% of the GA substrate remains in the granule core. The presence of other substances capable of PVCL complex formation in water, such as a resorcinol and tannin, results in a greater concentration release of GA from a granule.

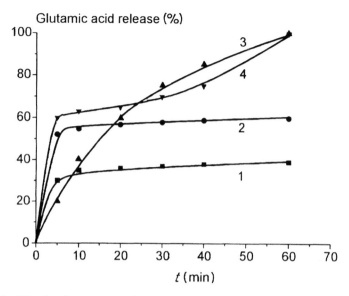

Fig. 4.15. Kinetic release curves of glutamic acid from PVCL formed granules depending on time of their formation, stabilizers: phloroglucinol-1, resorcinol-2, tannin-3 and sulfatrate containing polyamide PA-2 + resorcinol-4 [253].

In fact, the efficiency of PVCL precipitation on the introduction of phenols of various structures, determined by a turbidimetric titration as the minimum solution concentration of phenol (C_{min}) causing PVCL (0.7% in water) precipitation in an aqueous solution at the expense of their complex formation, is increased in the following series: pyrocatechol ($C_{min} = 3 \times 10^{-2}$ wt%) < phenol ($C_{min} = 1.6 \times 10^{-2}$ wt%) < hydroquinone ($C_{min} = 1.4 \times 10^{-2}$ wt%) < resorcinol ($C_{min} = 1 \times 10^{-2}$ wt%) < phloroglucinol ($C_{min} = 0.5 \times 10^{-2}$ wt%).

It follows from these data that the PVCL precipitation in water together with phenols depends on a structural arrangement of hydroxyl groups in a benzene ring. The most preferable interaction between polarized water molecules near to a PVCL chain and a phenol molecule is observed for resorcinol and phloroglucinol in which the groups are in an orthoposition in the benzene ring (see Section 4.3). The internal hydrogen bond between hydroxyl groups in pyrocatechol complicates their interaction with molecules of water near to a chain. The complex structure of tannin (polyphenols) requires more long duration time for the formation of a layer.

The cessation of the release of a low-molecular-weight substance from a granule (GA in fact not interacting with PVCL) demonstrates formation of a dense layer through which there is insufficient diffusion of small molecules, at least molecules of an investigated type (Fig. 4.15).

For formation of channels in this layer, in which in equilibrium condition diffusion of small molecules can occur, the aromatic water-soluble polyamide containing two sulfonic acid groups (PA-2, see section 4.4) is recommended as an additive stabilizer. The structure of the polymer chainlink is close to the structure of Blancphore dyes with two sulfonic acid groups (see section 4.3) which forms stable complexes with PVCL.

A looser layer generated from PVCL + PA-2 (Fig. 4.15) is permeable for low-molecular-weight substances: GA is completely released from a core. The introduction of PA-2 and resorcinol (0.05%) promotes the formation of a superficial dense layer with permeable channels. Such granules become steady at room temperature, and keep their form on long intensive agitation in water [254].

4.8 ENZYMES IN POLY-*N*-VINYLCAPROLACTAM SURROUNDING

It is known [255], that the catalytic activity of enzymes falls sharply with increase of temperature of a solution and in the presence of the various denaturating agents in an aqueous solution. The protein molecule of a enzyme is exposed to a denaturation with change of conformational state and loss of catalytic activity.

For the solution of a large number of biotechnological problems with use of enzymes as catalysts, there is the necessity for an increase of stability of enzymes in the sense that their catalytic activity should remain close to that of their native state at increase of temperature of a solution (> 37 °C) or in the presence of various chemical reagents. For the solution of this problem, a large number of studies with application to the various approaches of an immobilization [256] has been carried out.

Use of the PVCL macromolecule as a polymer matrix carrier to which the protein molecule is attached through a chemical bond (through a hydrolized caprolactam ring) [257], or as one of the gel particles of the polymer components in which the enzyme [254, 258] is included, essentially to raise the stability of enzymes to a thermoinactivation. Therefore, the process of a thermoinactivation of the chymotripsin enzyme appreciably decreases on the addition of this enzyme in conjugation with PVCL ($M_w = 6 \times 10^5$) in comparison with a native enzyme [257].

It is remarkable that the thermal processing of enzyme solutions (60 °C) during various times to a significantly lower degree deteriorates catalytic activity of the conjugate compared with the native one. It remains high at an exposure of 60 °C for two hours, and the activity of a native enzyme in reaction to an hydrolysis of an ethyl ether *N*-acetyl-γ-tyrosine disappears in the first 10 min treatment [257].

Such a conjugate of the PVCL enzyme is also interesting in that after realization of the reaction the enzyme, together with PVCL, is precipitated from a solution by heating to 40 °C, facilitating the separation process of the catalyst from products in the reaction medium. The cycle of thermoprecipitating–dissolving for enzymes immobilized on PVCL can be repeated without loss of catalytic activity of the enzyme [257], making these conjugates (PVCL–enzyme) attractive for biotechnology. The effect of PVCL on the stabilization of catalytic activity of a number of enzymes is also found in the composition of polymer hydrogels [258].

At present, for the preparation of polymer–enzyme gels, natural polymers, in particular polysaccharides and synthetic polymers, especially polyacrylamides, polyvinyl alcohols and others are used as matrixes [256]. However, the degree of enzyme incorporation in polymer gel and, accordingly, the relative enzymic activity of immobilized preparations are usually arrived at insufficiently high. Therefore, the activity of tripsin included in polyacrylamide gel is only 23% [259], and in hydrogel on the basis of polyvinyl alcohol, 70% [260]. Some enzymes such as carboxypeptidase cannot be included in gels because of their extreme instability in the native form.

With application of PVCL, new approaches of immobilization of such enzymes such as tripsin and carboxypeptidase were developed [258].

The activity of immobilized tripsin depends on the type and structure of a polymeric matrix, and also from the ratio of [PVCL]/[enzyme]. In a 'classical' alginate hydrogel the tripsin is not incorporated: only 7% of relative activity of the immobilized enzyme from the initial enzyme remains after immobilization in this gel. In PVCL gel tripsin is incorporated almost quantitatively (in the solution only 5–10% of the enzyme remained). Reactivity of an immobilized enzyme in PVCL gel 20–25% of the initial state.

The introduction in a PVCL matrix of alginate allows the activity of the enzyme to be increased, reduces the size of granules at the expense of a decrease of viscosity of a polymer solution and enhances the mechanical durability of the received gel [258].

For prevention of PVCL–enzyme complex diffusion from granules after temperature reduction, from 40 °C up to room temperature, granules are formed from PVCL with the sulfonate-containing aromatic polyamide (PA-2) which, as described above, forms a PVP–PA-2 complex. The great stability of enzymes is observed at PVCL, alginate and PA-2 concentrations being 2.5, 1.0 and 0.3 g in 100 g solution.

It was established [258] that in this system tripsin and carboxypeptidase maintained the catalytic activity during 4 months and 45 days respectively, at room temperature. At an optimum concentration (weight) ratio of [PVCL]/[enzyme] being 100 : 1, the relative activity of immobilized enzymes reached 85% (tripsin) and 45% (carboxypeptidase B).

It the enzyme is coupled to a copolymer of VP and diacetate acrolein by a covalent bond, such an enzyme included in PVCL–alginate–PA-2 gel retains the catalytic activity in a much greater degree (90% for tripsin and 80% for carboxypeptidase). It is essential that immobilization in such a gel does not change the optimum value of pH at which the maximum catalytic activity (pH = 7.5 for trypsin and 7.6 for carboxypeptidase B) is observed.

It is necessary to note that the PVCL macromolecules in gel bring in the basic contribution the increase of thermal stability: the enzymes retain relative catalytic activity up to 65–75 °C [258]. At the same time for tripsin stabilized by copolymer VP–acrolein complete inactivation occurs at 55 °C.

The granules with enzymes included were used for an enzymic hydrolysis recombinant proinsulin with the formation of insulin. A basic opportunity for the repeated use of the immobilized enzymes is shown in ref. [258].

On the basis of a urease and PVCL the stable immobilized preparations, effectively hydrolyzing urea [261], were received. These granules possessing a catalytic activity are formed by the addition of a solution of urease with PVCL ($M_w = 9 \times 10^5$) in a heated solution of resorcinol (45 °C). The data received in the study of catalytic activity at various pH and temperature of the immobilized urease testify that their catalytic properties do not differ from those of a native enzyme. The important characteristic of the enzyme in a PVCL granule in the fact that there is a significant increase in its stability at storage. Therefore, these catalytic systems retain their activity (90–100%) over six to eight months, while the native urease loses its activity during 14 days on 60%. In addition, the granular form of the preparation with enzymes is convenient for application in analytical purposes.

As the phenomenon of thermal denaturation of protein molecules and their stability in storage depend on a large number of factors and are not completely investigated [255], for an explanation of enzyme stabilization of the enzyme in PVCL surroundings it is necessary to use the knowledge of the characteristics of hydration and the conformational PVCL transformations in aqueous solutions on increase of temperature (see Chapter 3).

It was established by NMR methods, spin label, polarized luminescence and others, that at increase in temperature of an aqueous solution in PVCL macromolecules there are conformational transformations occurring at the shrinkage of macromolecules at the expense of hydrophobic and dispersion interactions of volumetric caprolactam chainlinks with partial replacement of water from a hydrate shell. At $T_{ph.s.}$ there is the aggregation of macromolecules with separation from the solution. Significantly, heating to 90–95 °C for a long time with subsequent decanting of water above a precipitate leaves a significant portion of water molecules in the precipitated polymer, namely about 12–14 molecules of water on one chainlink. In addition it was found (see Chapter 3), that the mobility of water molecules near to a PVCL chain becomes more restrained in comparison with water in volume.

The interaction of PVCL molecules with a protein molecule (enzyme) at heating seems to occurs principally through water molecules, being in hydrate shells of both PVCL and protein near to amide groups. The layer containing water molecules with restricted mobility forming a network of hydrogen bonds is created between molecules of protein and PVCL macromolecules retaining at heating up to higher temperatures (60–70 °C). PVCL macromolecules located close to active sites of protein molecules, in turn, can also restrict the effect of temperature on a conformational state of protein molecule fragments. The increase of temperature of a PVCL solution causes shrinkage of coils and is accompanied by a sharp delay in mobility of chainlinks. The PVCL macromolecules winding around protein molecules create the new construction surroundings from PVCL chainlinks and polarized water molecules which, with increase of temperature, become more and more rigid and as a result slows down the mobility of protein chains, preventing denaturation.

Thus, PVCL in an aqueous solution displays a unique property, representing itself as the stabilizer of the conformational state of protein against a thermal denaturation through an intermediate hydrate layer. In this intermediate water layer the water molecules form a network of hydrogen bonds with participation of both amide PVCL groups and amide and other functional groups (for example, COOH groups) of peptide chain. It is arranged on the surface of a protein molecule which it also protects from the denaturating action of organic additives such as dimethylformamide, glycerin and acetonitrile [258]. The presence of the hydrate layer between PVCL macromolecules and protein allows the creation of gel channels for diffusion of small molecules (substrates) and of protein molecule, for example proinsulin to active sites of immobilized enzymes without loss of their catalytic activity.

It is possible to include live cells of microorganisms in such type of gel, for example, *Gluconobacter oxydans, Corynebacterium glutamicum* and others [262, 263]. The precipitating phenomenon of the PVCL–enzyme conjugate from the reaction medium was also used for the isolation of tripsin interacting with a tripsin inhibitor on PVCL [264].

All this once again confirms that PVCL and the copolymers of VCL will find wide application in the creation of new processes in biotechnology, preparations in pharmaceutics and cosmetics.

4.9 POLY-*N*-VINYLPYRROLIDONE INTERACTION WITH A LIVE ORGANISM

Although a large number of published works and reviews [2, 70, 111] have been devoted to biological PVP activity in connection with its wide application in medicine it is interesting to discuss the reasons of such popularity of the given polymer.

PVP is used by the pharmaceutical industry in the manufacture of tablets and for a number of medical preparations since the 1940s. The approximate number of tablets manufactured in each year is about 100 billion [111]. Formulations of prolonged action with antibiotics, hormones and analgesics, preparations for ophthamology (contact lenses) and so on are produced on a base of PVP or VP copolymers [2, 43, 70].

In Russia intravenous preparations for detoxification (trade mark "Haemodesum") are widely applied. These preparations represents 6% of PVP in water–salt solutions (NaCl, KCl, CaCl$_2$) [2, 43, 196] and are produced in flasks of 100, 200 and 400 ml capacity. This preparation was developed and investigated in the 1960s and 1970s [43, 76, 196].

The Haemodesum preparation is widely used in clinical practice. A large number of works on the medical indications of this preparation have been published in Russia [265–286].

Transfusion of the preparation allows one to remove frustration of hemodinimics, of activity of gastrointestinal path, and of function of kidneys in result operating, burn and other kinds of traumas [265–270]. In the case of thermal traumas, an intravenous infusion of the preparation results in substantial growth of kidney circulation and glomerular filtration, to the cessation of vomiting, stimulation of activity of the lymphatic system, etc. [268, 271].

The physical form of children is improved on treatment of the toxic forms of gastrointestinal diseases of various a etiologies [272, 273]. It is recommended that the preparation is applied in complex treatments of bronchial asthma [274]. In combination with desensitizing preparations, it is supposed in the treatment of drug toxidermia [275] and for the oncologic sick [276].

In the treatment of psoriasis the Haemodesum preparation reduces the length of treatment with fast removal of skin. Application of the preparation avoids the application of corticosteroid hormones [277].

It was noted that rapid improvement of the somatic state occurred after introduction of the PVP solution with glucose (Neocompensan preparation in Austria) the sick suffering schizophrenia, alcohol psychoses with heavy abstinent syndrome [278], and also after introduction of Haemodesum to the sick with psychopathy state [279–281].

In a complex therapy of hepatitis [282] it is proposed to include detoxificating preparations of the Haemodesum type. The preparations have appeared effective in the treatment of chemical burns and eye diseases [283, 284].

Recently the issue of the preparation has been increased in Russia as it is recognized among clinicical physicians as an effective means of combating the causes of an intoxication of a various diseases, in particular thermal and traumatic fatalities, purulent-sepsis processes, pathology of the liver, toxic forms of acute gastrointestinal diseases, intestinal blackage, peritonitis virulent forms of pancreatitis, cholecystitis, appendicitis and others [196].

The medical efficiency of the preparation causes an improvement of microcirculation and liquidation of erythrocyte stasis in capillaries, that in turn leads to an improvement in kidney circulation and a sharp increase in diuresis. The preparation is entered as unitary, and repeatedly as a rate of infusions up to

10–15 times to one ill in a total dose of 2 or 6 l. The patients well transfer intravenous transfusions of the preparation; any reactions, including toxic and anaphilogenic are not marked.

The definition of blood toxicity and urine of sick people in mice before transfusion of the preparation reveals that up to an infusion of the preparation the blood has toxic action and the urine is not toxic. After Haemodesum infusion there is a sharp decrease of the toxic properties of the blood, while the urine becomes obviously toxic.

The medical efficiency of the blood substitute for detoxication action in toxaemias of a various origins is much higher than the detoxication property of integral blood [196].

The wide application of PVP in medicine could not be maintained if it had not a unique complex of physicochemical and biological properties, and also if estensive research were not undertaken, including study of its toxicological properties and interaction with a live organism (absorption, metabolism, distribution in bodies of rats, rabbits, monkeys and man). The results of these medical researches are incorporated as a monograph [111].

Research on animals and people confirm the fact that PVP is not a toxic substance. Therefore, the dose of LD_{50} at peroral and intravenous applications is 100 g/kg and 10–15 g/kg, respectively. It is not exposed to a metabolism in a live organism and its removal from an organism depends on MW and method of introduction.

At peroral introduction PVP absorption in an organism is minimum. All factions of PVP, except for oligomers ($M_n < 2000$), are not absorbed in the gastrointestinal path and are found in faeces.

At intravenous pharmocokinetics the administration of PVP elimination from an organism is related to MW of polymer. For example, after intravenous injection of labelled PVP with $M_w = 10 \times 10^3$ the reduction of a radioactive label in blood occurs through 15 min and its accumulation in the body is marked. The greatest amount of PVP is found in skin, liver, bone, kidneys, spleen, and lymph nodes.

Faster excretion of PVP in humans occurs in the case of PVP with a low MW. The PVP fractions with $M_w = 25 \times 10^3$ passes through glomerular capillaries and is found in urine. A 95 wt% of PVP with $M_w = 10 \times 10^3$ is eliminated from an organism during six hours. During 24 hours the elimination of 50 wt% of a PVP sample with $M_w = 25 \times 10^3$ proceeds from an organism. Hereinafter for 14 days 85 wt% of this polymer is eliminated.

Large macromolecules of PVP ($8 \times 10^4 - 1 \times 10^4$) do not penetrate through glomerular capillaries, having an average radius of pores of about 3.44 nm. Large polymer doses and high MW ($M_w = 5 \times 10^4$) polymer and the long-time of use of preparations with PVP (over several years) can cause undesirable storage-related functional changes in target organs [111].

Thus, numerous pharmacological researches of PVP performed during a long period (30 years) confirm the fact that at peroral introduction of PVP in a wide range of MW from 3×10^3 up 10^6 it is essentially non-toxic.

In addition, the ability of PVP macromolecules to form complexes with the most diverse substances, including biologically active substances in aqueous and water–organic solutions enables solid disperions with uniform distribution of the medicinal substance to be prepared. It also acts as a binding agent with solid fillers (talc, magnesium stearate and others).

It is usual for manufactured tablets to apply PVP with $M_w = 20 \times 10^3–30 \times 10^3$. Use of this polymer allows not only the preparation of tablets of high quality with properties required for medicine but also simplifies the technology of their manufacture.

That fact that the amide bond of rings is not exposed to hydrolysis because of structural restrictions in chain, unlike the low-molecular-weight analogue, explains the absence of chemical transformations of PVP in manufacturing solid dispersions, tablets and injection preparations which are sterilized by thermal processing and also in the biochemical actions of enzymes and other systems of living organisms.

A PVP macromolecule with a large hydrate layer near to chain does not cause gross disorder in the structure of cell membranes. The PVP macromolecules do not penetrate into cells and act, apparently, as a soft irritant of membranes and, as result, of cells.

The biological detoxication effect at intravenous administration of the aqueous 6% PVP solution seems to be due to the action of a large group of factors. It involves direct PVP interaction with toxins with the subsequent removal of a PVP–toxin complex through the kidneys, improvement of microcirculation of the blood, liquidation of erythrocyte stasis in capillaries, increase of diuresis, etc. However, the surprise phenomenon should be noted: that PVP treatment has a detoxication effect on toxaemias caused by toxins of various nature such as burn toxins, peritonitis, jaundice purulent infections and so on.

In experiments on animals at methanol poisoning the detoxication effect is also found on the introduction of PVP with various MW [285, 286].

All investigated solutions (6% PVP solution) which contain PVP with $M_w = 4.5 \times 10^3$, 15×10^3 or 46×10^3 display various physiological actions. The polymer solutions cause detoxication action, whereas 6% N-methylpyrrolidone solution reinforces the toxic action of methanol. The higher MW of PVP, the higher survival rate of animals after methanol toxaemia. Therefore, in the case of PVP with $M_w = 46 \times 10^3$ death of mice begins after 30 h, in the case of PVP with $M_w = 15 \times 10^3$, 8 h and in the case of PVP with $M_w = 4.5 \times 10^3$, 4–5 h. After 48 h, the survival rate is 70%, 47% and 35% for PVP with $M_w = 46 \times 10^3$, 15×10^3 and 4.5×10^3, accordingly, at a survival rate (26%) in the control group.

Further observation of the animals (5–7 days) has shown that despite the large number of deaths of mice in all groups, by 4–5 days the tendency to show the detoxication effect in greater degree is observed in polymers with high M_w.

Research on the biological activity of polymer solutions at various routes of administration has established that at intraperitoneal unitary injection of the solution, the detoxication effect is absent. At unitary intravenous injection this effect was poorly expressed. Only repeated fractional administration of solutions intravenously during 1.5–2 h results in a marked detoxication effect [285, 286].

Received results allow one to reveal the physiological action of PVP solutions on methanol poisoning which is connected to polymer MW: the higher MW in an investigated range (M_w = 4500–46000), the greater the number of surviving animals. Another important fact is that the toxic substance used (methanol) in an aqueous solution does not interact with PVP (see Chapter 3). It indicates the complicated mechanism of living organism detoxication with participation of PVP macromolecules which, apparently, cannot be attributed only to the formation of polymer + toxin complexes.

In this connection it is well to bear in mind reference data concerning the effect of MW of water-soluble polymers (PVP, PVA and dextrane) on an agglomeration of erythrocytes *in vivo* [287]. It shows appreciable interaction of PVP with an erythrocyte membrane surface. The degree of this agglomeration depends on MW: PVP with M_w = 20×10^3 in 25% concentration does not in practice cause an agglomeration, while PVP with M_w = 10×10^4 promotes the precipitation of erythrocytes. The radioactive method proves interaction of ^{14}C PVP (even the PVP with M_w = 20×10^3) with membranes of erythrocytes. It is essential that the basic role in this interaction is played by negatively charged lipids with phosphate groups.

Earlier it was established [288] that the intravenous PVP administration finds a reflection in the contents of prostaglandins and cyclic nucleotides in the spleen of mice at times from the beginning of the injection. Also, it testifies to a complicated character of PVP macromolecule interaction with a living organism, as the prostaglandins play a important role in the regulation of its immunity.

Other information on the action of PVP at a cell level was gained on studying the lipid composition of membranes of liver cells, as the basic organ participating in the detoxication of an organism, and being in a greater degree under the influence of toxic substances, is the liver [286].

The toxic influence of methanol on the lipid composition of the liver is greatly changed in comparison with livers of intact animals. The lipid composition of liver cell membranes in intact mice involves cardiolipin (CLP, 9.3%), phosphoditylethanolamine (PDE, 24.1%), phosphoditylcholine (PDC, 35%), sphingomyelin (SM, 9.7%) and the sum (11.6%) of two lipids from phosphoditylinositol (PI) and phosphoditylserine (PS), whereas after methanol intoxication for 24 h this composition becomes unlike: CLP, 13.6%; PDE, 42.2%; PDC, 33.1%; SM, 5.8% and PI + PS, 5.6%.

After the PVP administration (M_w = 15×10^3) it was displayed a change of the quantitative contents of basic lipid fractions (PHE, 30.8% and PDC, 40.5%) and the increase in contents of SM (14.2%) indicating partial normalization of lipid composition with some restoration of the properties of liver membranes.

The observable effect of the PVP operation on detoxication in the case of methanol, and the lipid composition normalization accompanying this effect indirectly confirm the hypothesis on the more complicated mechanism of detoxication than deactivation of a toxic substance by complex formation of a toxin molecule with PVP macromolecules. One can believe that PVP protection of a living organism against a toxin is connected to a certain degree to the strengthening of the protective function of membranes under the action of PVP [286].

This point of view, that cell membranes can be a target of attack by PVP macromolecules, is supported by data concerning the polymer action on composition lipids of liver membranes of intact animals. It turns out that the introduction of PVP increases a portion of easily oxidized lipids (PDE, 29.0%) by lowering the content of CLP (6.5%) and SM (6.4%) in comparison with the control group mentioned above.

In other words, the administration of PVP affects membranes of the liver: an authentic difference in the contents of a series of components of liver cell membranes after the injection of PVP (portion of cholesterol from 10.1% to 7.7%, portion of phospholipin from 1.0% to 5.1%) is observed. Therefore, the display of biological PVP detoxication effect cannot be explained only from the point of view of complex formation of toxic substances with PVP macromolecules, as for example methanol or products of its metabolism being uncombined with PVP in aqueous solution. Probably, in addition to the above-mentioned factor, additional factors promote the protective PVP action against toxin molecules.

The PVP macromolecules, through blood circulation, are quickly transferred to various sites of an organism and are absorbed in intercellular space. The weak "soft" interaction of the macromolecules with a surface of cells having a slight negative charge promotes the approach of these chains and fragments of membranes.

In Chapter 4 (section 4.1) it is stated that the PVP macromolecules, with the participation of polarized water in these hydrate shells are capable of the formation of weak complexes with inorganic anions, for example chloride anions, but not with cations. Therefore, the approach of PVP macromolecules to cell membranes can change the composition of an ionic atmosphere near to the membrane space at the expense of the redistribution of anions Cl^-. In turn, this is reflected in the activity of cells. Also, the effect of a specific hydrated shell near to the PVP chain on the near-surface aqueous layer of cell membranes seems to happen.

In the both cases PVP macromolecules, indirectly through interaction with cell membranes, strengthen the protective properties of the organism, reducing the adverse influence of toxins of the most diverse structure.

The further improvement in technology of PVP for medical purposes from the point of view of preparation of high-quality polymers (the more narrow MWD) in comparison with these at present ($M_w/M_n = 2.7–3.5$) can stimulate the future development of a new type of medicine with detoxication effect. It is possible that PVP with $M_w = 15\,000$ and $M_w/M_n = 1.3–1.5$, can be the basis for preparations of such a type as those will not essentially contain low-molecular-weight factions with MW $< 5 \times 10^3$ and high-molecular-weight factions with MW $> 30 \times 10^3–40 \times 10^3$. The former have a minimum ability for complexation, and the latter remains in an organism [76].

4.10 INTERACTION OF POLYMER–LOW-MOLECULAR-WEIGHT BIOLOGICALLY ACTIVE SUBSTANCE COMPLEXES WITH ORGANISM

PVP allows new medicinal forms, for example, injection preparations using biologically active substances poorly soluble in water to be created. In this case it favours the increase of solubility for these substances in water due to the complex formation reaction that makes it possible to realize the full physiological activity of a low-molecular-weight medicinal substance. Usually these preparations represent highly concentrated PVP solutions with medicinal compound for intramuscular introduction.

The use of aqueous 5–15% PVP solutions is recommended as a solvent for antibiotics, such as oxytetracycline, tetracycline, chlortetracycline and others [289–291]. The composition of 'Insipidine retard' preparation involves PVP with $M_w = 50 \times 10^3$ (20% concentration) and vasopressin [158].

Concentrated aqueous solutions with PVP dissolve molecules of medicinal substance because of the formation polymer complexes, the formation of these complexes being accompanied by essential changes in the biological action of this substance on a organism.

COMPLEX OF PVP + MORPHINE

In an example of the effect of PVP on the biological efficiency of a medicinal substance, being in a complex with PVP in an aqueous solution in an organism, a composition of PVP+morphine [292–296] can act. The value of the binding constant (K_b) between PVP and morphine should be low and in the region of 20–50 l/mol, as the morphine molecule contains two hydroxyl groups, one of which is connected to a benzene ring (hydroxyl of phenol type). The value of K_b is characteristic of phenols of various structures and dyes (see sections 4.1 and 4.2). This low value of K_b requires a high concentration ratio between PVP and morphine.

It turns out that in the study of the biological action of morphine in experiment with the application of sterile aqueous solutions containing various PVP concentrations and a constant concentration of morphine (0.5 wt%), the optimum PVP concentration was found to be 30 wt%. At smaller PVP concentrations the painkilling effect decreases [295–297]. In this case the high concentration ratio ($[PVP]/[morphine] = 180$) is achieved when all morphine molecules are bound by PVP macromolecules. The optimum value of MW which allows the preparation of concentrated solutions with a significant effect of painkilling prolongation is equalled $(30 \pm 5) \times 10^3$.

As is well known, morphine in an aqueous solution (1 wt%) is widely used in clinical practice as an analgesic for painkilling. However it has a number of disadvantages. One is the suppression of the cardio–vascular system, reducing heart rhythm at the expense of a single high concentration of morphine in blood. Another is that an occurrence of a nausea leading to vomiting is observed.

For a long-duration effect of painkilling, repeated administration (four to eight times) of morphine is required, creating a threat of dependence on the drug [292–294].

Zhorov and co-workers [293, 294] developed on the basis of PVP and morphine effective pharmaceutical composition for painkilling, which is documented in a pharmacopoeia article on 'Morphilongum'. This preparation (a 2 ml. ampoule from dark glass) represents a transparent viscous water solution of PVP and hydrochloride morphine sterilized at 120 °C for 20 min.

Let us consider the characteristics of the biological action of the PVP + morphine complex in comparison with a solution of morphine. First, pharmacokinetics of morphine in an organism at intramuscular administration of the preparation changes radically: in the case of free morphine its concentration in blood increases to 1.5–2.0 h; initially, with a subsequent appreciable fall of concentration to 8–10 h [293], while in the case of a complex the increase of its concentration in the blood is less (to 3–4 h), not reaching the maximum concentration which takes place on the introduction of free morphine [292–296].

Another important effect is that this concentration is maintained in the blood for 18–24 h by a slow fall of concentration in the subsequent 8–10 h. Secondly, clinical tests have shown that the painkilling effect lasts for 22–26 h in 70% of the patients, and in a number of cases for 32–34 h, depending on the illness.

That is why in the case of this preparation a basic analgesic effect is given, without narcotic action, unlike a pure morphine when the high concentration in blood creates both a painkilling effect and simultaneously, narcotic action. It is necessary also to note the absence of nausea in patients and less effect on the cardio-vascular system at administration of this preparation.

Therefore, the effect of habituation on morphine of 'Morphilongum', as have been shown in clinical tests (oncologic patients, postoperation and others) is not found. In addition, unitary intramuscular introduction of the preparation in amounts four to eight ml, depending on the weight of the patient, owing to a long-duration painkilling in four to eight times reduces number of injections for patients.

Thus, the PVP macromolecules at a certain concentration ratio with a low-molecular-weight medicinal substance in aqueous solution, and when they are in a complex, can affect the biological activity of this substance in an organism.

These effects are realized not only at the expense of the creation of a depot in the injection site with the subsequent slow elimination of polymer complexes in blood but also, probably, at the expense of a slowed-down diffusion of macromolecules of such MW holding the morphine molecule concentration in blood plasma. It was shown [111], that the PVP macromolecules ($M_w = 30 \times 10^3$) at intramuscular administration is eliminated from an injection site during one day on 86–89% of patients and observed in blood plasma and in lymph for three and more days.

COPYLMER + ANTISEPTIC COMPOUND COMPLEX

PVP or the copolymers on a VP basis containing the second monomer in the amount of 10–25 mol% in the majority of cases lowers the toxicity of low-molecular-weight biologically active substances, if the latter are in a complex with the polymer [299]. Also, such an example is the PVP–I_3^- complex in which toxicity of iodine is much reduced. The iodine in such a complex does not burn tissues and does not volatilize [43].

The mechanism of the toxicity decrease of small molecules in a complex with such a polymer is unclear, and in many cases depends on both a small molecular structure and the polymer structure. On the basis of the data in Chapter 3, one can assume that one of the primary factors defining the effect of reduction of toxicity of a small molecule in complex is the presence of a hydrate shell near to a macromolecular chain which surrounds the small molecule. Certainly, the other factor can be concentration of the substance in organism when depending on the binding constant by the polymer complex and following the elimination from an injection site.

Let us consider the characteristics of biological action of the VP–AA complex (or MAA or CA) copolymers with catamin AB, dimethylbenzylalkylammonium chloride with alkyl (from C_7 to C_{17}) substituents being effective antimicrobial agents [300–305]. The significant skin-irritating action of low-molecular-weight catamine-AB limits its use as local antiseptic.

By various physicochemical methods (turbidimetry, potentiometry [300–302], polarized luminescence method [237, 238] and others) it is shown that the formation of water-soluble complexes with the participation of copolymers of VP–AA (8.2–16.8 mol%), VP–MMA (10.7–29 mol%) or VP–CA (13.0–29.6 mol%) and catamine in aqueous and water–salt solutions. Thus the complexes are stabilized by both electrostatic interactions between charged groups on chain (COO⁻) and a positively charged ammonium group of catamine and hydrophobic interactions of apolar catamine groups. The stability of polycomplexes depends on pH, the ionic strength of a solution, the ratio of interacting components, structure and structure of the copolymer.

Microbiological tests [303–305] of polymer complexes of catamine—AB and the VP copolymers have shown that their direct antimicrobial activity depends on the complex stability in an aqueous solution: the more stable complexes are biologically less active. Therefore, for example, on the value of the minimum suppressing concentration (MSC) a more stable complex of catamine with copolymer of VP–AA (29 mol% of AA) has appeared two to four times less active than those with copolymer of VP–CA (15 mol%) containing the same amount of antiseptic.

Panarin with co-workers [301–305] have developed on the basis of a copolymer of VP–CA and catamine AB a medicinal preparation under the 'Catapol' trade mark.

It is necessary to note a significant reduction of toxicity of the polymer complex in comparison with catamine. The LD_{50} (mg/kg) value at intraperitoneal administration in mice increases from 22 ± 8 to 90 ± 19, and at subcutaneous introduction, from 47 ± 1 to 196 ± 7. This preparation has an appreciably smaller skin-irritating action and more expressed therapeutic effect at treatment of purulent wounds and burns. At present the 'Catapol' preparation is used in veterinary surgery for hand washing of veterinary personnel, for treatment of freshly infected wounds and preventive maintenance of purulent postoperation complications in cattle.

5

Reactions in Poly-*N*-Vinylamide Chains

5.1 REACTION OF ACIDIC HYDROLYSIS

The involvement of a functional group in a polymer chain significantly affects its reactivity compared to the involvement of the former in a low-molecular-weight analogue [191, 306]. Revealing the factors causing this polymer effect is of great importance for understanding the mechanism of reactions in chains such as complex formation, hydrolysis and others.

The fact of high hydrolytic stability of PVP in water in comparison with that of a low-molecular-weight analogue is well known [2]. Only prolonged boiling with acid or alkaline brings about partial PVP hydrolysis with formation of chainlinks of *N*-vinylaminobutyric acid [307, 308]. The opening of 80–90% PVP rings can be carried out only in very severe conditions in a melting cryoscopic mixture of NaOH–KOH at a temperature of 280 °C [309].

Due to this property of PVP its aqueous solutions endure thermal processing needed for medicinal preparation sterilization (110–135 °C) without changes in PVP structure [2].

The attempt to elucidate the factors of PVP hydrolytic stability in water was made using poly-*N*-vinylamides of various structure, where substituents at both the nitrogen atom and the carbonyl group are altered [310]. It is found that heating of a polymer aqueous solution with HCl (1 N) at 100 °C during 60 h causes complete hydrolysis of poly-*N*-vinylformamide and poly-*N*-vinylacetamide (100%) and insignificant hydrolysis of PVP (4.2% hydrolysed chainlinks) and of PVCL (1.5%).

The structure of poly-*N*-vinyl-*N*-methylformamide and poly-*N*-vinyl-*N*-methylacetamide remains virtually unchanged, indicating a very high stability of amide groups with methyl substituents at nitrogen atom (reaction 5 1):

$$\left[\begin{array}{c} -CH_2-CH- \\ | \\ N \\ CH_2 \diagdown \diagup C=O \\ CH_2-CH_2 \end{array}\right] + H_2O \xrightarrow{H^+} \left[\begin{array}{c} -CH_2-CH- \\ | \\ N \\ CH_2 \diagdown \diagup C=O \\ CH_2-CH_2 \end{array}\right] 95.8\% \left[\begin{array}{c} -CH_2-CH- \\ | \\ NH \\ | \\ (CH_2)_3 \\ | \\ COOH \end{array}\right] 4.2\%$$

$$(5.1)$$

Poly-*N*-vinylformamide hydrolysis proceeds to poly-*N*-vinylamine and formic acid [89, 90] (reaction 5.2):

$$\left[\begin{array}{c} CH_2-CH \\ | \\ N \\ H \diagup \diagdown C=O \\ | \\ H \end{array}\right] \xrightarrow{H^+(H_2O)} \left[\begin{array}{c} CH_2-CH \\ | \\ N \\ H \diagup \diagdown H \end{array}\right]_n + HC\diagup\diagdown^O_{OH} \qquad (5.2)$$

It is known [114] that catalysis of the amide hydrolysis reaction in an acidic medium is possible with formation of an intermediate, where proton is bound to the oxygen atom of C=O (reaction 5.3):

$$R_1-\overset{\overset{O}{\|}}{C}-NH-R_2 + H^+ \longrightarrow R_1-\overset{\overset{OH}{|}}{C}=N^+H-R_2 \qquad (5.3)$$

It was shown in an example of *N,N*-dimethylformamide by the ^{1}H NMR method that the C=O group of amide protonates in an acidic medium [46]. Similarly, the carbonyl group of PVP interacts with proton [310] (Fig. 5.1). In a range of pH≈1.0 the protonization of C=O with an increase of partial positive charge on the carbon atom (Fig. 5.1) takes place. An appreciable displacement of chemical shift of the carbon atom in $-C_xH-$ (hetero- and syndio-triads) also occurs, together with a weak displacement of this shift in iso-triads. It indicates the differing abilities of lactam rings of chainlinks, in these configurations, to interact with proton.

Proton interaction with a pyrrolidone ring in PVP chainlink can be presented as follows:

$$\begin{array}{c} -CH_2-CH- \\ | \\ N \\ CH_2 \diagdown \diagup C=O \cdots H^+ \\ CH_2 - CH_2 \end{array} \quad \begin{array}{c} H \\ O\diagdown \\ \diagdown H \\ O\diagup^H \\ H \end{array}$$

Comparing the reactivity of the amide group in poly-*N*-vinylamides (PVP, PVCL, PVMA and PVMF) in an acidic hydrolysis reaction one can state that retardation of the hydrolysis reaction is subjected to the greatest influence in

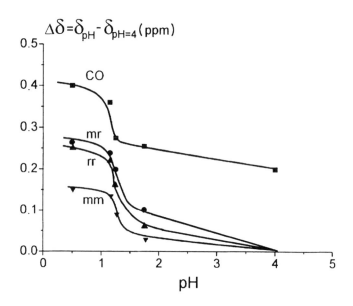

Fig. 5.1. The effect of pH of solution on chemical shift of carbon atoms in C=O and C_xH— groups in a ^{13}C NMR spectrum of PVP [310]. Reproduced by permission of Nauka.

the case when the nitrogen atom bound to a methine group of chain has an additional methyl group. In the case of the replacement of the methyl group by a hydrogen atom, hydrolysis of the amide group proceeds quickly and completely.

This different hydrolytic reactivity of polymer amides with various substituents proves significant of the structural factor in realization of this reaction in terms of certain coordination of water molecules near to the amide group [114].

The course of this reaction is stipulated by an arrangement of atoms where the oxygen atom of water is near to the carbon atom of the C=O group and the proton of water molecule is near to the nitrogen atom. The size of the X group at nitrogen on a chain defines this orientation. The methyl group (or methylene groups of a ring) probably lessens the coordination of the water molecule near to the amide group due to steric hindrance. The replacement of CH_3— by H cancels this effect.

A very small fraction of PVP or PVCL chainlinks participate in the hydrolysis reaction which seems to be due to structural distortion in an arrangement of chainlinks of greater size, resulting in the accessibility of only a few chainlinks to interaction with the H_2O molecule being in the above-mentioned coordination relative to the amide group.

It should be noted that the absence of interaction between water molecules and the nitrogen atom of PVP was also proved by the IR-spectroscopic method [120]. This charateristic of hydration of amide groups of a polymer, such as PVP or PVCL, is common for all poly-*N*-vinylamides with *N*-alkyl substituent.

5.2 MODIFICATION REACTIONS OF POLY-*N*-VINYL-AMIDES CONTAINING REACTIVE GROUPS

ANTHRACENE-CONTAINING POLYMERS

The expedient selection of the chemical structure of homo- and copolymers of *N*-vinylamides in order to solve various practical problems and optimize conditions of their application requires the study of intramolecular mobility and structure formation of these polymers in solutions (see Chapter 3 and 4). It is also necessary to carry out research on the molecular mechanism of interactions between polymers of various structures and low-molecular-weight compounds and to analyse the factors stabilizing or destroying complexes of polymer and small molecules. Such investigations, carried out at a molecular level, require methods which allow the study of polymers in very dilute solutions (down to 0.01%) and in various solvents or their mixtures at different pH values, ionic strengths, etc. In addition, if the research on intermolecular interactions is carried out, one should be provided with an opportunity to study each of the components of a multicomponent polymer system.

The method of polarized luminescence developed by Anufrieva and co-workers [248, 306] corresponds to all these requirements. However, this method demands the application of fluorophor-labelled polymers where fluorophor (luminescence) groups are attached by covalent bond to a polymer chain (the label amount is usually 0.1 mol%).

It was established [311, 312, 313] that among a set of luminescent groups which can be attached to polymer chains as luminescent labels, the 9-alkylanthracene group has a special advantage. Krakoviak and co-workers [312, 313, 314] proposed different approaches to the introduction of this fluorophor in polymers of various structures, particularly in poly-*N*-vinylamides and copolymers of *N*-vinylamides.

Copolymers with carboxylic groups interact with 9-anthranyldiazomethane in the following way (reaction 5.4):

$$\text{(5.4)}$$

9- anthranyldiazomethane

The reaction of 9-anthranyldiazomethane with COOH groups on chains of VP copolymers proceeds quickly and in mild conditions (at low reagent concentration, at room temperature and at the absence of catalysts) [312, 313]. One can use this reaction for quantitative determination of carboxylic groups at very low contents (< 0.1%) in polymers [314].

Copolymers of VP, VCL and VMA with acrylic, methacrylic and other unsaturated carboxylic acids react with 9-antranyldiazomethane [312], as well as the partially hydrolysed chainlink in PVP, namely *N*-vinyl-aminobutyric acid (reaction 5.5):

$$\text{(5.5)}$$

Copolymers containing hydroxyl or amine groups easily interact with the isocyanate derivative of anthracene. It was established [314, 315, 316] that 9-anthranylmethylisocyanate is a suitable reagent for the addition to various amino- or hydroxyl-containing copolymers of *N*-vinylamides.

This reagent is easily added to copolymers of *N*-vinylamides such as VP–vinylamine (VA) or allylalcohol (AA) (reaction 5.6). It reacts with terminal

$$(5.6)$$

hydroxyl groups of PVP ($M_w = 1 \times 10^4$) obtained in the traditional way (VP, H_2O, H_2O_2, NH_3) allowing PVP macromolecules to be synthesized with a luminescent label on a terminal chainlink.

VP copolymers containing aldehyde groups can interact with 9-anthranylamine forming a Schiff base [312]:

$$(5.7)$$

Anazomethine bond formed on the reaction of the addition of 9-antranylamine to the aldehyde group is stable in an aqueous solution of copolymer for at least a few days. Processing of a polymer system with azomethine groups by boronhydride results in the formation of stable luminescent labels on a polymer chain.

The realization of the reactions of VP–acroleine copolymer (10 mol% of acrolein chainlinks) specified above gives a fluorophor-labelled copolymer containing 0.16 mol% chainlinks with an anthracene group (reaction 5.7).

It is important that anthracene-containing copolymers of N-vinylamides are water soluble and their application as a label could help to solve diverse scientific problems of biology, biotechnology and medicine.

COPOLYMERS CONTAINING p-NITROPHENYL ESTER RESIDUES

VP copolymers with *p*-nitrophenyl ester groups are of interest as a basis of the preparation of a large group of water-soluble biologically active polymers. These copolymers allow one to solve a problem of creation of a number of medicinal substances with long-term prolonged effects.

Basic requirements of these copolymers are a certain MW, narrow MW distribution and quick release from an organism after fulfilment of their biological function by renal filtration, similar to PVP removal.

Solovsky, Panarin and co-workers [317–321] developed in detail various pathways of synthesis of such reactive polymers on the basis of copolymers, obtained with the use of VP as the basic monomer, and their subsequent transformation into biologically active products. It is essential that VP chainlinks in these copolymers help to maintain solubility of copolymers in water. Such copolymers can contain an appreciable amount of chainlinks of another monomer (up to 10–25 mol%) bearing reactive groups capable of subsequent chemical transformations.

Therefore, one has to consider a series of copolymers on a VP basis containing reactive *p*-nitrophenyl ester groups [317, 318]. Copolymers of VP with acrylic (AA), methaclylic (MAA), crotonic (CA) and *p*-acryloylaminophenoxyacetic (AAPA) acids are used for the addition of nitrophenyl ester residues. *N*,*N*-dicyclohexylcarbodiimide (DCC) acts as a condensation agent in esterification reaction between polymer acids and *p*-nitrophenol.

The degree of esterification of VP–AA or VP–CA copolymers in the investigated reaction was found to achieve 70%, while in the case of VP–MAA and VP–AAPA copolymers it was 36%.

The results of research [319, 320] of the aminolysis reaction of *p*-nitrophenyl ester residues in the obtained copolymers testify that high reactivity is characteristic of ester groups in the copolymer of VP–AAPA due to distance from the main chain. They can enter the reaction with sterically hindered primary aminocompounds (*tert*-butylamine or ampicillin antibiotic). Ester side residues in copolymers of VP–AA or VP–CA are able to react with non-branched primary amines, e.g. (3,4-dihydroxyphenyl)ethylamine, forming amide groups on a chain. The fraction of groups entering this reaction is 33–35%. Water-soluble polymer derivatives of (3,4-dihydroxyphenyl)ethylamine contain 2.4–5.1 mol% of bound biogenic amine [321].

5.3   IONIZATION EQUILIBRIUM OF POLYMER AMINE GROUPS

Understanding of the behaviour of ionic groups, separated from each other on a poly-*N*-vinylamide chain, in a reaction of ionization equilibrium, allows one to explain the character of local surroundings near to a polymer chain in aqueous solution. The investigated polymer systems, with small amine group contents on a chain surrounded by polar amide groups, can serve as models of slightly charged protein molecules.

Copolymers of *N*-vinylamides with vinylamine (VA) (VMA–VA, VP–VA and VCL–VA) and allylamine (AA) (VMA–AA, VP–AA and VCL–AA) with low amine content were subjected to test [322].

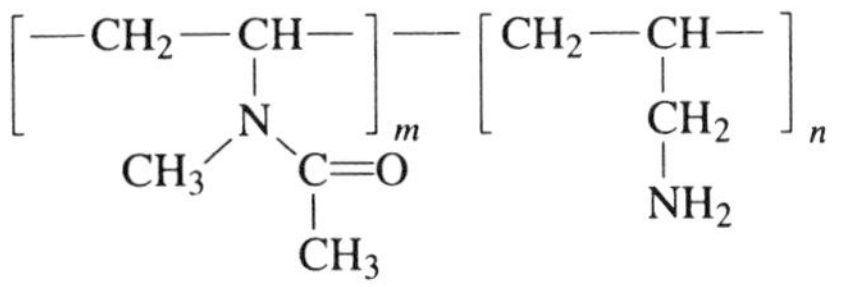

$$\left[-CH_2-CH- \right]_m - \left[CH_2-CH- \right]_n$$

with

$$\begin{array}{c} N \\ CH_3 \diagdown \diagdown C=O \\ | \\ CH_3 \end{array} \qquad \begin{array}{c} CH_2 \\ | \\ NH_2 \end{array}$$

CP–VMA–AA (*n* = 0.25, 2.8, 10 mol%)

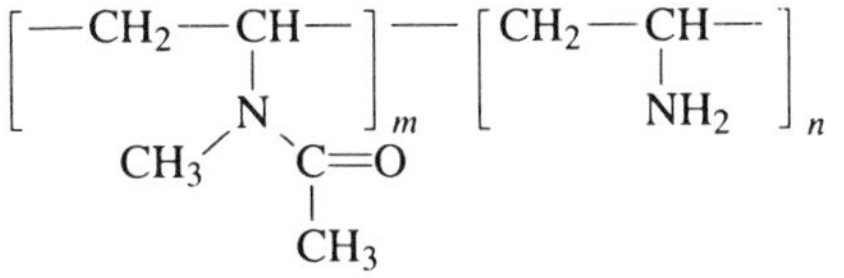

$$\left[-CH_2-CH- \right]_m - \left[CH_2-CH- \right]_n$$

$$\begin{array}{c} N \\ CH_3 \diagdown \diagdown C=O \\ | \\ CH_3 \end{array} \qquad \begin{array}{c} | \\ NH_2 \end{array}$$

CP–VMA–VA (*n* = 1.9, 4.3, 7.0 mol%)

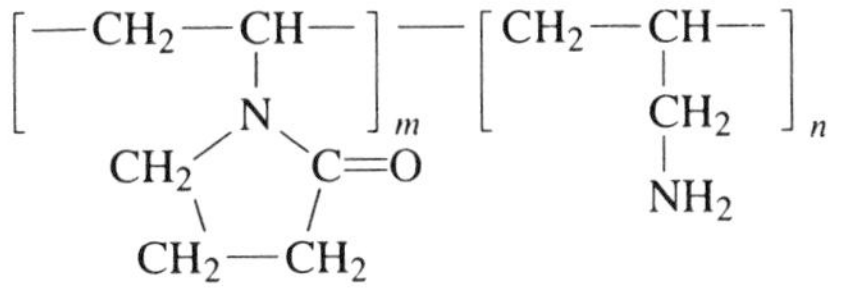

$$\left[-CH_2-CH- \right]_m - \left[CH_2-CH- \right]_n$$

$$\begin{array}{c} N \\ CH_2 \diagdown \diagdown C=O \\ \diagdown \diagup \\ CH_2-CH_2 \end{array} \qquad \begin{array}{c} CH_2 \\ | \\ NH_2 \end{array}$$

CP–VP–AA (2.5 mol%)

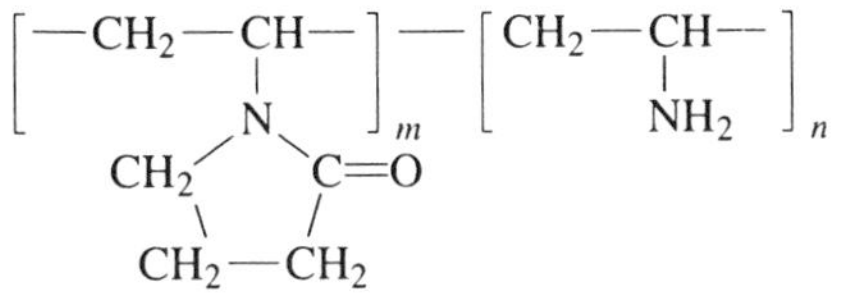

$$\left[-CH_2-CH- \right]_m - \left[CH_2-CH- \right]_n$$

$$\begin{array}{c} N \\ CH_2 \diagdown \diagdown C=O \\ \diagdown \diagup \\ CH_2-CH_2 \end{array} \qquad \begin{array}{c} | \\ NH_2 \end{array}$$

CP–VP–VA (*n* = 3.3, 8.0 mol%)

$$\left[-CH_2-CH- \right]_m - \left[CH_2-CH- \right]_n$$

$$\begin{array}{c} N \\ CH_2 \diagdown \diagdown C=O \\ \diagup \qquad \diagdown \\ CH_2 \qquad CH_2 \\ \diagdown \qquad \diagup \\ CH_2-CH_2 \end{array} \qquad \begin{array}{c} CH_2 \\ | \\ NH_2 \end{array}$$

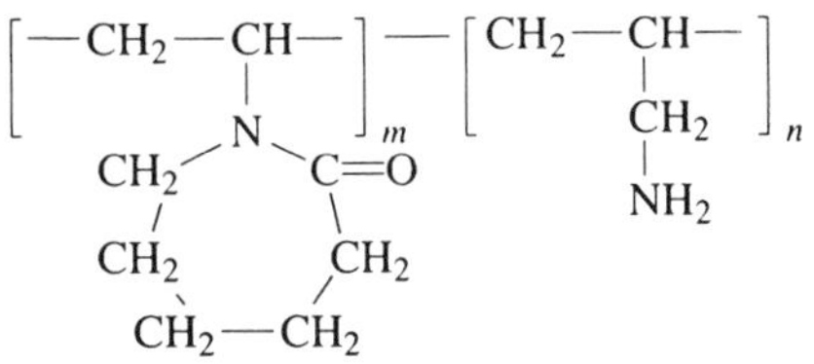

CP–VCL–AA (*n* = 3.7 mol%)

$$\left[\begin{array}{c} -CH_2-CH- \\ | \\ N \\ CH_2 \diagdown C=O \\ CH_2 \quad CH_2 \\ CH_2-CH_2 \end{array}\right]_m \left[\begin{array}{c} CH_2-CH- \\ | \\ NH_2 \end{array}\right]_n$$

CP–VCL–VA (n = 1.7; 3.8 mol%)

The analysis of dependencies $pK_{a(\alpha)}^{\text{eff}} = pH - \log\alpha/(1-\alpha)$ (α is the degree of ionization of amine groups) for all investigated copolymers (Fig. 5.2) allows one to reveal a number of characteristics of the behaviour of amine groups in an acid–base equilibrium.

First, the values of pK_a^0 of amine groups of all investigated copolymers obtained by an extrapolation of $pK_{a(\alpha)}^{\text{eff}}$ at $\alpha \to 0$ are much lower than the pK_a of

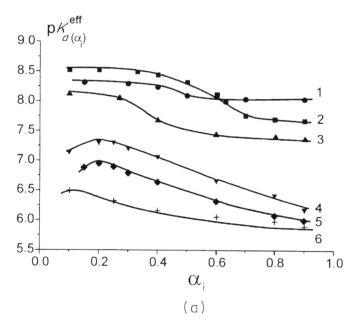

(a)

Fig. 5.2. The dependence of $pK_{a(\alpha)}^{\text{eff}}$ of amine groups on ionization degree (α): in copolymers of VMA (a): CP–VMA–AA containing 10 mol% (1), 2.8 mol% (2) and 0.25 mol% (3) of AA, and in CP–VMA–VA containing 7.0 mol% (4), 4.3 mol% (5) and 1.9 mol% (6) of VA; in copolymers of VP (b): CP–VP–AA containing 2.5 mol% of AA (1) and CP–VP–VA containing 8.0 mol% (2) and 3.3 mol% (3) of VA; in copolymers of VCL (c) CP–VCL–AA containing 3.7 mol% (1) and CP–VCL–VA containing 3.8 mol% (2) and 1.7 mol% (3) of VA [322]. Reproduced by permission of Nauka.

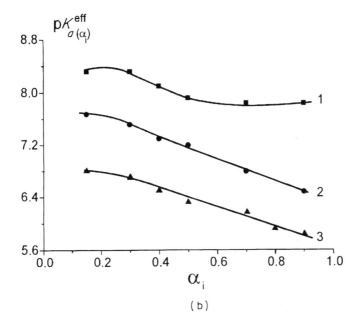

(b)

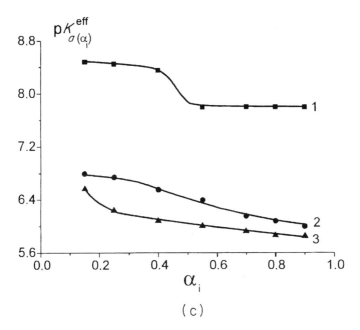

(c)

Fig. 5.2. (*continued*)

low-molecular-weight analogues of chainlinks with the amine group. In fact, the values of pK_a^0 of VMA–AA, VP–AA and VCL–AA copolymers are equal to approximately 8.5, 8.3 and 8.4 respectively, in an investigated range of amine group contents; pK_a^0 values for copolymers of VMA–VA, VP–VAA and VCL–VA are in the range 6.6–7.2, 7.0–7.8 and 6.7–6.9 respectively. At the same time pK_a for chainlink analogues (isobutylamine and isopropylamine) is 10.4 and 10.6 respectively [99].

Hence, at the small contents of amine groups in a chain of copolymer of N-vinylamide and allylamine or vinylamine the groups become weak bases. The value of the dissociation constant (K_a) for copolymers with allylamine exceeds that for the analogue 100 times, and that for copolymers with vinylamine 1000 times. It is interesting that this effect is observed for amine groups arranged far from each other on a chain (20–100 chainlinks apart) in the absence of positive charges created by protonic amine groups.

Comparison of plots of $pK_{a(\alpha)}^{\mathrm{eff}}$ vs α for copolymers with various side substituents and various arrangements of amine groups relative to the main chain allows one to reveal the effect of the nearest surroundings on the ability of the amine group to interact with proton. In copolymers where the side residue varies from methyl-acetamide group to pyrrolidone ring and caprolactam ring, the value of pK_a^0 differs only insignificantly. At the same time as the amine group approaches the main chain, the pK_a^0 value decreases considerably, being 6.6–7.8 for copolymers with VA (amine is bound to the main chain) and 8.3–8.5 for copolymers with AA (amine is separated from the main chain by a methylene group).

The reduction of the pK_a value of amine groups in copolymers is caused by the fact that these groups are located on a chain in close proximity to partially positively charged carbon atoms of C=O groups (see Chapter 1). As negatively charged oxygen atoms of C=O dipoles are hydrated by water molecules and positive charges are inaccessible for solvent molecules, these positive charges in slightly polar areas probably reinforce the electrostatic fields, preventing the approach of proton to an amine group.

The ability of amine groups to interact with proton seems to be regulated by its arrangement relative to a positively charged end of C=O dipoles. This arrangement can vary due to structural heterogeneity of the main chain (config-uration isomerism) resulting in the occurrence of amine groups with different basicity.

In the case of VMA–VA, VP–VA and VCL–VA copolymers, monotonous decrease of $pK_{a(\alpha)}^{\mathrm{eff}}$ with the increase of α cannot be explained by the effect of positive charges appearing on a chain, as the distance between amine groups is significant (50–200 chainlinks) hence they would be titrated independently. At higher contents of amine groups (e.g. in copolymer VMA–AA with 10 mol% of AA) the value of pK_a^{eff} is virtually independent of α.

For VMA–AA, VP–AA and VCL–AA copolymers where amine groups are also separated from one another by a large number of amide-containing chain-links, the relation curves of $pK_{a(\alpha)}^{\mathrm{eff}}$ vs α are sigma-like.

Most likely, the sigma-like character of $pK_{a(\alpha)}^{eff}$ vs α relation curves for all AA-containing copolymers and monotonous decrease of $pK_{a(\alpha)}^{eff}$ with the increase of α for VA-containing copolymers is stipulated by different micro-surroundings of the amine groups, being affected by both configurational chain heterogeneity of amine arrangement (syndio-, iso-and heterotriads) and conformational isomerism of a side residue (allyl substituent).

Amine groups of AA-containing copolymers are separated from a chain and hence from dipoles of the neighbouring amide-containing chainlinks. The action of partial positive charges located close to the amine group on the ability of the latter to interact with proton is defined both by the distance between the amine group and dipoles and by local permittivity near to a chain. The latter, in turn, depends in a complex manner on the conformational isomerism of allylamine residue.

These factors probably define the sigma-like character of the dependence curve of $pK_{a(\alpha)}^{eff}$ vs α. Amine groups in syndiotactic triads in AA-containing copolymer chain, being titrated first of all, possess the greatest basicity (see $pK_{a(\alpha)}^{eff}$ vs α curve in a range $0 < \alpha < 0.25$ (^{13}C NMR shows that the percentage of syndiotriade fraction in the PVP chain is 24%). Slight lowering of $pK_{a(\alpha)}^{eff}$ vs α curve takes place in a range $0.25 < \alpha < 0.6$–0.8 that can be stipulated by ionization properties of amine groups arranged in heterotactic sequences (43% of triads in atactic PVP chain). Finally the lowest values of $pK_{a(\alpha)}^{eff}$ are observed in a range of 0.6–$0.8 < \alpha < 1$ where amine groups in isotactic triads are titrated. The fraction of isotactic triads is 33% (by ^{13}C NMR).

The flattened curves of $pK_{a(\alpha)}^{eff}$ vs α for copolymers of VMA–VA, VP–VA and VCL–VA where titratable sites are separated from each other by large distances, seem to be due to the slight effect of structural arrangement of amine groups, being in this or that triad, on ionization properties of the groups. In addition, these amine groups are located close to a chain and are subjected to the action of both close and remote $C{=}O$ charges at protonization.

The occurrence of functional groups of various reactivity in synthetic polymers was found earlier in copolymers of vinylpyridines with acroleinoxime or vinylthiol [24]. In the first case, where the functional groups (oxime groups) are well off the main chain, a large set of groups of largely different nucleophility were revealed, whereas in the second case the occurrence of such sites was in a smaller degree.

Typical curves of $pK_{a(\alpha)}^{eff}$ vs α for copolymers of VMA–AA and VMA–VA in the presence of 0.5 mol/l of KCl, KBr and KI solutions are given in Figs. 5.3 and 5.4.

A similar picture was observed for the other copolymers with AA and VA. Such character of the dependencies testifies to the weak effect of ionic strength on $pK_{a(\alpha)}^{eff}$. Usually the increase of ionic strength in solutions of the other polyelectrolytes increases $pK_{a(\alpha)}^{eff}$ of ionic groups of alkaline type, suppressing electrostatic effects. In this case the pK_a^{eff} of copolymers with AA and VA even reduces, indicating the absence of electrostatic interaction between remote amine groups.

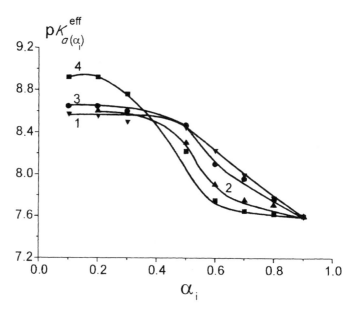

Fig. 5.3. Dependence of $pK_{a(\alpha)}^{eff}$ on ionization degree (α) CP–VMA–AA (AA content is 2.8%) in the absence (1) and the presence of KCl (2), KBr (3) and KI (4) [322]. Reproduced by permission of Nauka.

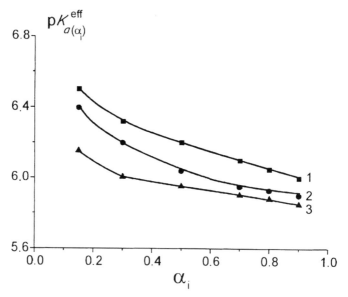

Fig. 5.4. Dependence of $pK_{a(\alpha)}^{eff}$ on ionization degree (α) CP–VMA–VA (VA content is 1.9%) in the absence (1) and the presence of KCl (2) and KI (3) [322]. Reproduced by permission of Nauka.

Some influence of salts on the basicity of amine groups could be seen through their action on conformational and hydration states of macromolecules. The addition of salts, especially KI in solutions of copolymers with AA, causes a slight increase of basicity of the amine group being in a syndiotactic triad and a slight decrease of basicity of amine groups being in other triads. It is possible that amine groups being in syndiotactic triads are the most accessible to anions promoting some increase of iodide concentration near to amine groups and the attraction of protons.

Thus, the behaviour of amine groups in copolymers of *N*-vinylamides with AA or VA in small contents testifies to the presence of amine groups of different reactivity due to heterogeneity of their micro-surroundings (hydration, conformational state and configurational structure) in a polymer chain.

To some extent functional (amine) groups of such copolymers can serve as models of ionic groups of protein molecules which are characterized by significant variation of basicity and acidity in comparison to low-molecular-weight analogues due to micro-heterogeneity of their local surrounding [323].

5.4 POLY-*N*-VINYLPYRROLIDONE + IODINE REACTIONS IN SOLUTION AND IN POWDER

AQUEOUS SOLUTIONS

It is known that iodine is poorly soluble in water (0.3% or 1.2×10^{-2} mol/l). In water iodine is exposed to hydrolysis with very low $K_{dis} = 5 \times 10^{-13}$ mol/l [321].

$$I_2 + H_2O \leftrightarrow HI + HOI \qquad K_{dis} = [HI] \cdot [HOI]/[I_2] \qquad (5.8)$$

Such a low value of K_{dis} testifies that the equilibrium in reaction (5.8) is shifted to the left.

The introduction of PVP in water with iodine crystals and the subsequent lengthy shaking of a mixture shifts the equilibrium to the right and causes other reactions [2, 204, 206, 325]. The effect of PVP ([PVP] > [I_2]) is expressed in the complex formation of PVP macromolecules with I^- appearing as a result of I_2 hydrolysis (see Chapter 4) that accelerates hydrolysis iodine reaction (reaction 5.8) [325]. As a result the rate of solid iodine dissolution is increased as iodine interacts with I^- forming triiodide complexes (reaction 5.9).

$$I_2 + I^- = I_3^- \qquad (5.9)$$

Then PVP macromolecules bind I_3^- at a rather high constant of stability ($>10^4$ l/mol [208]). This results in the shift of equilibrium reaction (5.8), increasing hypoiodous acid concentration (HOI) followed by its conversion to iodic acid (reaction 5.10):

$$3\,HOI \leftrightarrow 2\,HI + HIO_3 \qquad (5.10)$$

As result of the conversion reaction of I_2 in water in the presence of PVP, the iodine concentration is changed with the increase of solution pH.

Varying crystal iodine concentration (0.0262, 0.035, 0.044, 0.065 and 0.075 mol/l) at constant PVP concentration in water (0.9 mol/l) at protected shaking (about one day) one can prepare an iodine solution of concentration equal to 0.26, 0.4, 0.52, 0.7 and 1.0 wt% respectively [325]. Iodide anions appear in solutions. In a series of these solutions iodide concentration becomes 0.0108, 0.0148, 0.017, 0.025 and 0.031 mol/l, and pH values are 1.97, 1.85, 1.79, 1.70 and 1.68 respectively. A strong acidic medium is created because of the presence of HI and HIO_3 acids.

Thus the interactions of PVP macromolecules accelerate the iodine hydrolysis reaction rate. About 30–40 wt% of added iodine is converted to iodide and IO_3^- (reaction 5.10). The conversion reaction (5.11) does not pass completely, and a new equilibrium is established in the presence of PVP:

$$3I_2 + 3H_2O \leftrightarrow 5HI + HIO_3 \qquad (5.11)$$

The value of solution pH is defined by both HI and HIO_3 concentrations and by acidic dissociation constants of HI and HIO_3 ($K_{diss} = 2 \times 10^{-1}$ mol/l). The remaining 60–70 wt% of iodine reacts with HI forming I_3^- (reaction 5.12) which in turn complexes with PVP.

$$I_2 + HI \leftrightarrow HI_3 \qquad (5.12)$$

PVP POWDER + CRYSTAL IODINE

An understanding of the mechanism of iodine + PVP conversion reactions in aqueous solution allows one to suggest a mechanism of similar reactions occurring at 'dry' mixing of powder-like PVP and crystal iodine.

The 'dry' method of obtaining PVP–iodine complexes for medicine (mechanical stirring of PVP powder and crystal iodine) was proposed for the first time by Shelanski [204]. This approach possibly remains the basic process in the manufacture of antiseptic products based on PVP + iodine complexes. Extensive information on other synthetic methods of PVP + iodine preparation in solutions (water or organic solvent) is also available in the patent literature [43].

The PVP + iodine complex (red-brown powder) is manufactured by BASF (Germany) and ISP (USA) as Povidone Iodine™. In Russia it is produced under the trade mark Iodopyronium. This powder containing 9–11 wt% of iodine is the basis of a large number of medicinal antiseptic formulations (ointments with polyethyleneoxides, aqueous iodine solutions (1 wt%), shampoos and other medicinal forms) [43]. Iodine in these forms does not volatilize and does not burn tissue.

The preparation process of a finely divided powder-like complex of PVP and crystal iodine is described in numerous patents (see review [43]) and includes stirring of powder-like PVP and finely divided crystal iodine for five hours at room temperature. Then the mixture is heated to 70–90 °C and kept at this temperature for 10 hours. The contents of iodine titrated by thiosulfate is from 9.0 to 11.0 wt%.

As PVP powder, even after industrial spray drying, contains up to 5 wt% water (it can also be sorbed from atmosphere), the same reactions occur during mechanical stirring as in solution.

The first reaction (5.8) with the participation of extrinsic water promotes the formation of HI, decreasing initial iodine concentration by 30–40 wt%. The acids (HI and HIO_3) formed in PVP + iodine powder dissociate due to the strong interaction of protons with C=O of amide groups, resulting in the formation of positively charged polymer chains (see reaction 5.13).

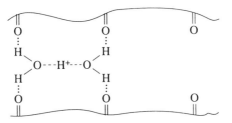

$$ (5.13) $$

This reaction seems to be one of the basic reactions, allowing the distributions of crystal iodine in PVP powder almost uniformly. Fast H^+ diffusion (basically $H_5O_2^+$) in 'channels' among PVP chains with the participation of C=O groups at stirring and heating (reaction 5.13) provides penetration of H^+ followed by I_3^- into hollow areas in powder polymer.

Formation of I_3^- in a solid matrix occurs due to the equilibrium reaction (5.9) where the equilibrium is shifted almost completely to the right. The fraction of iodine (60–70 wt%) which has not reacted with H_2O is bound by I^- forming I_3^-.

Since the molar concentration of I^- formed at reaction (5.9) is less than equimolar (15–25%) in relation to $[I_2]$, the occurrence of an additional reaction (5.14) is possible:

$$ I_3^- + I_2 \leftrightarrow I_5^- \tag{5.14} $$

The dissolving of crystal iodine in powder-like PVP at stirring and heating is due to certain physicochemical factors.

The major factor is $H_5O_2^+$ diffusion in areas of high local concentration of polar amide groups of high dipole moment (~ 4.0 D). In addition, significant electrostatic fields created by C=O dipoles weaken the interaction of water molecules with C=O on separate chainlinks in $H_5O_2^+$ transfer 'channels'.

The concept is based on studying of water molecule self-diffusion in a film of aromatic polyamide by NMR with magnetic field gradient method [326]. Self-diffusion coefficient of water in such matrix at very small water contents is rather high, being 15–20 times less than that of bulk water. High mobility of water molecules is attributed to the C=O effect in transfer 'channels', weakening water molecule interaction with the matrix and with other water molecules.

I_3^- ions, despite their large size, penetrate through the 'channels' created by H_2O and $H_5O_2^+$ molecules. The interaction between I_2 and C=O groups also makes a contribution to the dissolving of finely divided iodine in PVP powder (see section 4.2) [204, 206]. The finely divided red-brown powder attained consists of PVP + I_2 + HI + HIO_3 components interacting with each other. Aqueous solution of this powder (at 1% iodine) has an acidic reaction at pH = 1.5–1.7 (basically 1.5). That is why it is necessary to introduce neutralizing agents (e.g. trisodiumphosphate) to avoid tissue irritation. Such solution with various additives (in particular surface-active substances) is widely used in medicine.

5.5 STYRENE HETEROPHASE POLYMERIZATION

The presence of poly-*N*-vinylamides of various structures was found to affect heterophase radical polymerization of styrene (St) in water, initiated by potassium persulfate [327]. The effect is displayed in the variation of all polymerization process parameters such as reaction rate, size of particles, polydispersion coefficient, coagulum concentration, aggregation resistance in water–salt solution and at storage.

The presence of PVP, PVMA, poly-*N*-vinyl-*N*-ethylacetamide (PVEA) and poly-*N*-vinyl-*N*-propylacetamide (PVPA) in polymerization reaction medium at certain conditions (70 °C, volume ratio of $[St]/[H_2O] = 1:9$, concentration of $K_2S_2O_5$ and polymer equal to 1 wt% of a mixture) affects the reaction rate, being 0.3%, 0.2%, 0.28%, 0.16% of St per minute respectively. Styrene particles of average diameter equal to 0.30 (PVP), 0.35 (PVMA), 0.35 (PVEA) and 0.65 (PVPA) mkm at a polydispersity coefficient equal to 1.01 were obtained after the performance of the reaction.

It should be noted that styrene suspensions prepared in the presence of investigated polymers possess narrower size distribution of particles and contain no coagulum. The suspensions are stable in 0.3 mol/l NaCl solutions and at storage during three months. The investigated polymers appeared to be the best stabilizers compared to traditional polyvinyl-alcohol (PVA) stabilizer, although their surface tensions on the water/*o*-xylene boundary are quite similar (PVA 20.0, PVP 20.7, PVPA 21.5, PVEA 23.0 and PVMA 22 mN/m).

Styrene suspensions obtained in the presence of PVA contain coagulum (5%) and have wide particle size distribution (polydispersity coefficient is equal to 1.45) [327].

The effect of PVMA concentration on the initial polymerization reaction rate (V_0), the average diameter (d) of polystyrene particles and their polydispersity coefficient is of great interest. On the increase of PVMA concentration in a series 0.1wt%, 0.25, 1.0, 2.0 and 4.0 wt%, V_0 increases from 0.05% to 0.15%, 0.28%, 0.41% and 0.62% of St per minute respectively, while the d value of particles diminishes from 0.84 μm to 0.78, 0.36, 0.26 and 0.20 μm in the same sequence of PVMA concentration at low polydispersity coefficient (1.03–1.01).

One of the peculiarities of suspension radical polymerization of styrene in water is that styrene droplets in water on intensive stirring in the absence or in the presence of polymer have an average diameter of about 10 μm. After the introduction of an initiator ($K_2S_2O_8$) the styrene polymerization process begins, together with fast conversion of large monomer drops into monomer–polymer droplets of small size (0.2–0.6 μm). It is essential that these small droplets appear at ~10% conversion of monomer in the presence of any poly-N-vinylamides studied. Droplet size is defined by polymer concentration and structure. A monomer–polymer droplet formed at the beginning of the polymerization reaction retains its average size and dispersity (1.01–1.04) almost to the completion of the reaction. It is remarkable that the polymerization reaction produces polystyrene suspension particles of 0.3–0.35 μm in the presence of PVMA and 0.5–0.6 μm in the presence of PVPA at the same concentration of the added polymers.

Hence poly-N-vinylamide stabilizer except PVCL promoting coagulum formation regulates particle size, prevents the suspension from coagulation and makes possible the retaining of the monomer–polymer droplet size until complete conversion of monomer inside them. As the transformation of large drops into small occurs at the beginning of the polymerization reaction, one can assume that it is caused by various reactions proceeding on a drop surface [327].

$$^-O-\overset{\overset{\displaystyle O}{\|}}{\underset{\underset{\displaystyle O}{\|}}{S}}-O-CH_2-\overset{\bullet}{C}H \quad \overset{\left[\begin{array}{c}CH_2\!=\!CH\\ \end{array}\right]_n}{\longrightarrow}$$

$$^-O-\overset{\overset{\displaystyle O}{\|}}{\underset{\underset{\displaystyle O}{\|}}{S}}-O-CH_2-CH-[CH_2-CH]_{n-2}-CH_2-CH-$$

The persulfate radical formed approaches the styrene molecule located on a drop surface and attacks the double bond of styrene, starting a chain propagation inside the large monomer drop. The terminal sulfate group hydrated by water molecules remains on the surface. The sulfate-containing styrene chainlink is accessible to the complexation reaction with poly-N-vinylamide macromolecules.

It is well known that there are many examples of sulfonate-containing aromatic compounds interacting with poly-N-vinylamide macromolecules in aqueous solutions, as is shown in Chapter 4. As a result, terminal groups of polystyrene chains remaining in water and being on a drop surface promote the attachment of poly-N-vinylamide fragments to a drop surface due to complex formation.

The findings concerning the stabilizer concentration effect on the size of suspension particles support this suggestion. In fact, different polymer concentration favours the formation of different size suspension particles: low concentration (0.1–0.5 wt%) promotes the formation of large size particles (0.7–0.8 mkm), whereas at 1–4 wt% smaller ones (0.2–0.35 μm) are formed.

Macromolecules, surrounding drops, create a water–polymer layer preventing the coalescence between various monomer–polymer droplets and, thus the aggregation between polystyrene suspension particles and promoting the suspension stability in aqueous solutions.

There is a limited number of sites on a drop surface where the complex formation with the water-soluble polymers inestigated is possible. Naturally the greater the macromolecule concentration in a solution, the larger the number of areas on a drop surface near to which hydrate polymer chains are situated. It is possible that the number of such areas also makes a contribution to the transition of large monomer drops (10 mkm) into small monomer–polymer droplets, as reduction of the drop size takes place with the increase of polymer concentration [31].

Other physical factors causing this transformation at the beginning of polymerization reaction (at < 10 wt% conversion) could possibly be the high rate constant of chain propagation reaction (145 1/mols), a considerable thermal effect (75.3 kJ/mol) and a very low coefficient of heat conductivity (12.5×10^{-4} J/cm s °C) of styrene liquid.

In the sites on a drop surface where initiation and styrene chain propagation reaction start, it is possible that localized overheating can arise due to poor heat conductivity of the styrene medium, resulting in thermal explosion of a large monomer drop and formation of small droplets, where the heat is removed more effectively into the atmosphere [327].

Thus, styrene polymerization reaction in water can be controlled by the introduction of effective stabilizers of a new type, namely, poly-*N*-vinylamides of a certain structure. It is remarkable that these polymers possessing low surface tension affect heterophase polymerization of both styrene and other monomers (methylmethacrylate, chloroprene and others) regulating particle size, together with size distribution and increasing suspension resistance in water–salt solutions. The sorbed poly-*N*-vinylamide macromolecules carrying water shells are responsible for the increased resistance of suspension in water, preventing coagulation of suspension particles. This statement is supported by the fact that polystyrene suspension particles obtained in the presence of PVMA possess increased sorption ability towards protein molecules compared to those obtained in the absence of PVMA.

5.6　PVP EFFECT ON ULTRAFILTRATION MEMBRANE FORMATION

Recently poly-*N*-vinylamides, particularly PVP, attract attention of researchers in the field of chemistry and physicochemistry of membranes and membrane processes. The research in this sphere is developed rather actively at the moment as a large number of technological processes with the use of membranes are applied in the chemical, petrochemical and food industries. Some problems of maintenance of human vital activity concerning water treatment can be solved only with the use of membranes [328, 329].

PVP is used as an additive in chemically resistant ultrafiltration (UF) membranes formation on the basis of polymers of various structures, e.g. polyacrylonitrile (PAN) [330], polysulfone (PS) [331, 332], polyphenylsulfone [333], polyethersulfone (PES) [334, 335], polyamide-4,6 [336] and others.

In the manufacture of asymmetric UF membranes, highly concentrated (15–30 wt%) polymer solutions in amide solvents (DMF, *N*-methylpyrrolidone, dimethylacetamide or others) are prepared. Then a layer of viscous liquid of a certain thickness is cast on a support (e.g. on a glass plate) and immersed in cooled water (precipitator) [337, 338].

During the precipitation of polymer in a system of two liquids (a solvent (DMF or DMAM) and a precipitator (water)) the pores are formed. The pore sizes, size distribution and pore morphology (pore depth asymmetry) depend on many factors including interactions: polymer–solvent, polymer–precipitator and solvent–precipitator [337]. These interactions are adjusted by the polymer concentration, the humidity of an atmosphere where cast film is contained, the temperature of precipitator, the time of exposure of a film in precipitator, the concentration of additives, the temperature of drying and other conditions.

It turns out [332–336] that PVP introduction markedly affects the morphology of pores. The polymer reduces the formation of macropores, permitting ultrafilters of narrower size distribution to be obtained. It also creates conditions for the formation of membranes of high porosity of the upper layer and sublayer of membrane.

Asymmetric hollow-fibre ultrafiltration membranes are obtained from the polymer composition containing 15 wt% of PVP ($M_w = 1.5 \times 10^4$) and 28% of PS in dimethylacetamide (DMA) [339]. PVP contributes to the carring out of this method, because it forms homogeneous solutions with PS in DMA, being a water-soluble polymer.

As in ref. shown [336], the introduction of PVD affects properties of the polymer solution during membrane formation by the method of immersing in the precipitator.

During the precipitation of polyamide (polyamide-4,6) macromolecules under the action of water, PVP macromolecules cause deterioration of solubility of polymer material in a water-organic system. A large amount of PVP greatly affects the concentration ratio of components (polyamide–PVP–water–DMA) at which the system turbidity occurs (precipitation process) [336, 340]. The data on variation of light transmittance of solution cast films during precipitation in the presence and in the absence of PVP testify that PVP reduces the induction time since the start of immersion until the occurrence of turbidity of the polymer layer in the precipitator (water) [336].

Thus one could assume that a significant increase of solution viscosity caused by PVP introduction could result in lowering of the precipitator–solvent exchange rate and in the increase of induction time. However, it is remarkable that PVP of high MW ($M_w = 3.6 \times 10^5$) (5% concentration) [336], induces instantaneous precipitation of the basic polymer which forms a membrane. The smaller amount and MW of PVP ($M_w = 2.4 \times 10^4$ or 4×10^4) also diminish the induction time in comparison with that of the initial solution without PVP.

The explanation of this phenomenon is proposed in refs. [336, 340]. It stems from the fact that during the precipitation of polyamide-4, 6 and other polymers mentioned above (PS, PES and PAN) PVP shows itself as a good precipitator.

Such an unusual role for PVP which forms homogeneous solutions together with a basic polymer in amide solvent, but becomes a precipitator at the immersion of this polymer (PS, PES and PAN) in water, could be attributed to the special qualities of the hydrate surroundings of PVP chainlinks (see Chapter 3).

First, it is known [2] that only dry PVP is dissolved in $CHCl_3$, CH_2Cl_2 and dichlorethane. PVP loses this ability if it contains even a small amount of water

(5–10 wt% relative to PVP). In this case PVP is soluble only in water. This PVP property is used in the industrial process of extraction of a residual VP monomer by CH_2Cl_2 from PVP aqueous solution [2,69].

As a result, at the immersion of a viscous polymer cast layer containing PVP in water, different conditions of aggregation of basic polymer macromolecules (polyamide 4,6 or PS) are created.

PVP macromolecules, uniformly distributed among macromolecules of precipitating polymer make a contribution to the formation of a porous system from the gel-like layer at immersion in water. Probably, the creation of the upper layer of high porosity without macropores and uniform distribution of pores on a surface is caused by participation of PVP macromolecules in several processes accompanying the process of phase separation of the basic polymer.

During water penetration into polymer gel, PVP macromolecules provide uniform transfer of water molecules in the depth of this layer and their even distribution. Water molecules quickly penetrate between macromolecules of matrix polymer with the help of PVP amide groups, acting as 'bridges'. Water molecule associates penetrating deeply into the layer (basic polymer + PVP + DMAA) enter the hydrate shell near to PVP chains and come under the influence of carbonyl dipoles. Hydrate shell around PVP chainlinks containing water molecules polarized by C=O dipoles becomes a poorer solvent for a membrane forming polymer than bulk water. It accelerates the process of phase separation as is established in refs. [334, 336, 340].

5.7 SELECTIVITY AND WATER TRANSFER THROUGH REVERSE OSMOSIS MEMBRANES BASED ON AROMATIC POLYAMIDES

Some VP copolymers (with vinylacetate, ethylacrylate or methylmethacrylate) were used in the manufacture of reverse-osmosis (RO) membranes [341, 342]. Thin films made of industrial PVP samples modified by polyisocyanate possess salt rejection ability and high water permeability under high pressure [341].

RO membranes made of these materials are not applied in practice as they do not conform to a large number of the requirements of selectivity of a barrier layer and water permeability and also to their chemical and thermal resistance. However, the fact that salt rejection by thin layers of copolymers containing PVP chainlinks was detected, testifies to a certain role of amide groups in the functioning RO membranes.

In this section the operation mechanism of RO membranes is considered using new data obtained by studying poly-*N*-vinylamide hydration (see Chapter 3).

The point is that considerable progress has been made in the production RO membranes, while many scientific problems are still waiting to be solved [343]. Insufficiently developed are the concepts of selectivity and permeability of membranes influenced by the chemical structure of polymers, the processes occurring during membrane formation, the nature of functional groups in the chain and their mutual orientation, and other factors. Therefore the further development of the new types of RO membranes is restrained by the lack of knowledge about the RO membrane operation mechanism.

In order to solve the problem a thorough understanding is required of physicochemical phenomena concerning the structure of water associates near the surface of a selective layer, interactions of water molecules with functional groups being on the surface and inside the layer, the role of functional groups both in water transfer through the layer and in the separation of salt ions and water molecules at the entrance in polymer layer, self-diffusion of water molecules in channels, channel size, and so on.

Let us consider the structure of polymer materials used at selective layer formation, in order to reveal the basic types of functional groups favouring water transfer.

It should be noted that the list of polymer materials used for RO membrane formation is rather long. A large number of publications deals with polymer materials used as bases of RO membranes [343–350]. A wide spectrum of polymers of diverse structure is used in the preparation of a barrier (selective) layer of asymmetric or composite RO membranes, such as aromatic polyamides, polyethers, sulfonate-containing polysulfone, heterocyclic polymers, e.g. polybenzimidazole, polyoxadiazole, polyquinozoline and polymers of other types.

In the case of asymmetric RO membranes, as they are known [345–348], the barier layer is formed by phase inversion of polymer at the immersion of a viscous solution of cast film in water as a precipitator. In the same way asymmetric RO membranes were prepared from aromatic polyamide (m-phenylendiamine (MPD) and a mixture of isophthaloyl and terephthaloyl chlorides ($7:3$)) (99.8% salt rejection and 10 $1/m^2$ h permeability for NaCl 3.5% solution at 10 MPa) [347], and also from heterocyclic polymer specified above [349].

In the case of composite membranes, active components (low-molecular-weight or high-molecular-weight substances) are applied to a support surface (ultrafilter made of PS) with the subsequent thermal processing. A surface layer could be obtained, for example, by the therrnal polycondensation of furyl alcohol and polyethylene oxide with sulfuric acid [346] (performance of the membrane: 99.9% salt rejection and 33 $1/m^2$ h for 3.5% NaCl solution at 6.9 MPa), by interphase polycondensation of aromatic acid diclorides and aromatic or aliphatic diamines (polyethyleneimine + isophthaloyl chloride polymer, the membrane performance: 99.0%, 30 $1/m^2$ h for 3.5% NaCl solution at 10 MPa [50]), 1,3,5-benzenetricarboxylic acid chloride + MPD (FT-30 membrane performance: 99.2%, 40 $1/m^2$ h for 3.5% NaCl solution at 5.5 MPa [346]).

The analysis of RO membrane references allows one to conclude that the polymers investigated have a common feature of chemical structure, namely, high molar fraction of groups or atoms capable of hydration binding with water molecules. Such groups are amide ones in aromatic polyamides, ether groups ($-CH_2-O-CH_2-$) in condensation polymers, sulfonate and carboxylic groups or nitrogen atoms in heterocyclic polymers.

Further, in this section no polymer materials and performance results of RO membranes are examined because many reviews covering this information are availabe [343–350]. Instead special attention is focused on processes and reactions forming a barrier layer, on functional groups present in a matrix and on structural transformations of water associates near to a layer surface and inside it.

A common feature of chemical reactions forming a material of a selective layer of 100–200 nm in thickness, e.g. of a composite membrane, is the fact that the elimination of low-molecular-weight product (H_2O molecules) takes place at thermal polycondensation [346], and the elimination of HCl molecules at inter-phase polycondensation with simultaneous formation of intermolecular bonds (crosslinking). In such processes the elimination of small molecules promotes the formation of pores (channels) in a layer, and chain crosslinking fixes their size. Thus a rigid skeleton is created with cavities (or pores) filled with water, providing system stability and reducing local movement of chainlinks [343].

Therefore the synthesized polymer material of the FT-30 membrane barrier layer, as is shown by the ESCA method [351], contains 72% of crosslinked and 28% of branched forms. The polymer structure can be presented as follows:

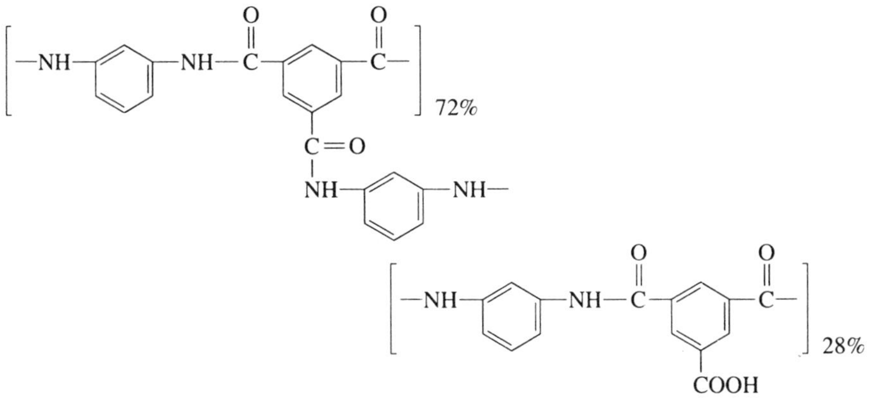

Where the material does not contain crosslinkings, the selectivity of a layer is sharply reduced [346]. It is essential that the layer prepared of trimellitic anhydride acid chloride and MPD possess a reduced selectivity (60–80% against 99.2% salt rejection for FT-30):

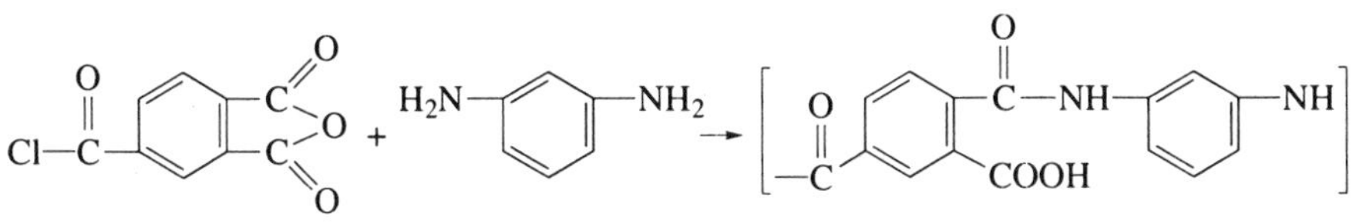

The reaction system in which crosslinking is impossible (1,3,5-benzenetri-carboxylic acid dichloride + MPD) also shows poor selectivity after layer formation:

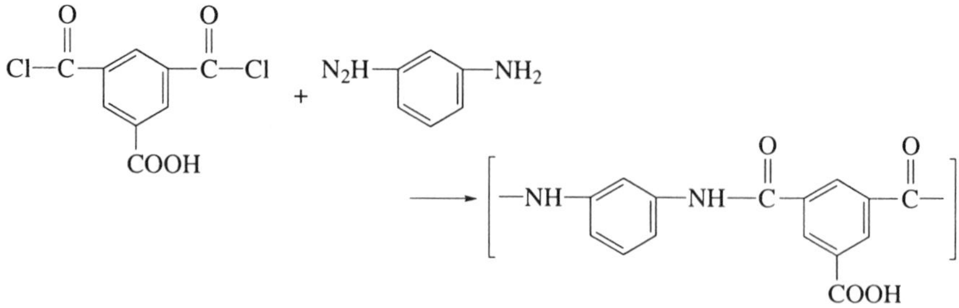

In asymmetric RO membranes the pores (or channels) are formed at the immersion of partially dried film into cold water [347]. However, at phase separation an opportunity still exists to achieve equilibrium conformational rearrangement of chains that could reduce the general porosity of a selective layer. Therefore, comparison of the performance of asymmetric and composite membranes [352] (5.6 MPa, 3.5% NaCl solution, 25 °C) allows us to state that all known asymmetric membranes have two to four times less water permeability than composite ones. During polycondensation the crosslinking process of formed chains and the elimination of small molecules 'pushing' apart polymer chains and creating channels inside the thin layer occur simultaneously.

Recently some progress has been achieved in studying qualities of ion and water transport and hydration in films made of aromatic polyamides (sulfonate-containing polyphenylene isophthalamides) having a structure close to that of aromatic polyamides of the selective layer [353–356].

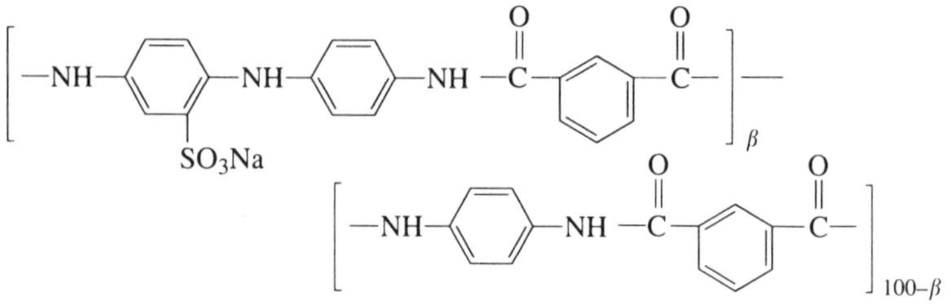

CPA–1

The rigid design of channels, uniform distribution and water contents in matrix control channel section sizes in matrix and stipulate selective transport of various size cations and of various charge cations under the action of electric current [353, 355]. The value of specific electroconductivity, as was found [353, 355], depends on the nature of the cation and is increased in the following series: $Ca^{+2} \ll Li^+ \ll Cs^+ < K^+ < Na^+ < H^+$.

Only at $N < 5$ (water molecule number per polymer chainlink) the selectivity of cation penetration appears: the mobility of larger size cations (Cs^+) in the channels is delayed more than that of the smaller Na^+ (by the radioactive isotope method) [353,356].

It is essential that polyamide films showing selectivity of cation transport under the action of electric current are characterized by the high transfer number of Na^+ ($t_+ = 0.95-0.99$) and weak sensitivity to the decrease of this number at the increase of salt concentration (up to 3 mol/l). It means that the channels in a film (30–40 μm thick) are resistant to the action of salt of high concentration and virtually do not pass through chloride anions. However, at $N = 6-7$ the cation transfer number decreases considerably, indicating that the channel section size increases and chloride anions penetrate inside.

Thus, varying a ratio of chainlinks of two types (containing and not containing a sulfonate group) in CPA-1 polyamide allows one to simulate in thicker films the characteristic parameters of a thin selective layer of RO membrane made of aromatic polyamides (e.g. FT-30 [346], FT-30BW [357] or Aramide B-9 [358] RO membranes): channel size, hydration degree influencing this size, chain rigidity and others.

Assessment of the average size (diameter) of the section of a transport channel with water, organised in cation-exchange membranes made of polyamldes specified above ($\beta = 30-40$ mol% of sulfonate-containing fragments), characterized by high selectivity an appreciable specific electroconductivity gives 8.5–9.5 Å [350, 353, 359].

It is important to note that the Na^+ transfer number in the films investigated, obtained from aromatic CPA-1 polyamide ($\beta = 30-40$ mol%), almost coincides with the NaCl salt rejection coefficient (0.98–0.992) for composite and asymmetric RO membranes made of aromatic polyamides. It means that in an RO membrane selective layer the channels are also formed of a rigid design and of similar size, impermeable to chloride anions [356].

Low mobility of atoms of polyamide chains in the two crosslinked and two linear polyamides prepared via interfacial polycondensation of phenylene diamines (*m*-and *p*-phenylene diamines (MPD, PPD)) with either tri-functional (trimesoyl chloride (TMC)) or di-functional (terephthaloyl or isophthaloyl chloride (TPC, IPC)) compounds was established by the cross-polarization/magic angle spinning (CP/MAS) ^{13}C NMR method [360]. The performance (selectivity and permeability) of RO membranes made of these polymers was correlated with their chain mobility. On the basis of the measurement data of spin-lattice relaxation time in the rotating frame ($T_{1\rho}$) it was concluded that mobility plays an important role in the regulation of water permeability through the active layer irrespective of the presence of crosslinks: the less mobile the chains, the higher the water flow.

The atomic force microscopy method (AFM) was used to investigate morphology and properties of the surfaces of RO thin-film-composite (TFC) membranes made of four polyamides [360]. MPD/IPC and PPD/TPC membranes show fairly similar morphology of grainy structure with nodule agglomerates. Tri-function acid chloride at polymerization gives a surface of well-developed 'ridge-and-valley' structure. An IPC + MPD polymer selective layer of longer T_{lp} relaxation time and grainy nodule surface is characterized by higher water permeability, whereas the TPC + PPD polymer layer of shorter T_{lp} relaxation time and rougher surface ('ridge-and-valley' structure) possesses higher salt rejection.

As the dimensions of the observed features of surface structure of polyamide active layers are about 10 Å, and the diameter of channel sections does not exceed 8.5–9.5 Å, these features can be considered as factors controlling the separation process of water molecules and salt ions. The average size (about 9.0 Å) of the diameter of the channel section where water transport is used is slightly different from the difference in heights of 'ridge-and-valley'. Probably the roughness of the surface where a large number of carbonyl groups is located affects ionic distribution near to channel entrances and increases selectivity.

Let us discuss the role of other factors defining the operation mechanism of a selective aromatic polyamide layer using the facts obtained on the study of poly-*N*-vinylamide hydration (see Chapter 3). These factors could be the interaction of water molecules with C=O dipoles surrounded by apolar groups and structural transformations of water associates, accompanying this interaction.

Carbonyls of amide groups together with COO^- or SO_3^- groups (in smaller amounts) are located on the channel entrance walls and inside the channel where water molecules move. These dipoles seem to affect water molecules near to the layer surface and in intra-channel space in a different way.

One can assume that negatively charged ends of C=O dipoles located on the entrance walls of the channel form hydrogen bonds with water molecules and polarize the nearest surrounding water, influencing the interaction between water molecules and the structural organization of water associates located close to the channel entrance. Considerable transformations of water associates in hydrate shells near to polymer chains of high local concentration of C=O amide groups (poly-*N*-vinylamides) and to PVCL gels are shown by the DSC method (see Chapter 3).

Specific structural organization of water molecules in bulk water is also a fundamental factor of water separation from a salt solution in the channels of selective layer at RO.

As was shown by X-ray analysis method [172], water is a strongly associated liquid. It contains associates—clusters where a central H_2O molecule is surrounded by 4.4 molecules, the distance between molecule centres being 2.9 Å, i.e. the first hydrate layer. The distance between the centres of water molecules of the second hydrate layer and the central one is 4.9–5.0 Å.

One can assume that the C=O of amide groups located on channel entrance walls (as well as C=O carboxylic or S=O sulfonic acid groups of smaller dipole moments) participate in co-operative interaction with a water associate of four to five molecules because of hydrogen-bond formation and is in fact the second solvate layer. These $C=O$ groups are located at the same distance from the centre as the second hydrate layer (about 4.5–5.0 Å). The rigid design of the channel containing water maintains these sizes and defines the transfer selectivity of water molecules.

The entry of cations into the channels of such a size is complicated due to the large size of hydrate cation shell: cation dehydration requires a large expenditure of energy at the entrance to the channel. It is known [328], that in a series of cations: Li^+ ($r = 0.78$ Å), Na^+ (0.93 Å), K^+ (1.33 Å) and Cs^+ (1.65 Å) [324] at the identical charge, salt rejection falls although cation size is increased. This effect is caused by the gradual reduction of cation hydration energy in the above-mentioned series. In fact, the values of the hydration heat of Li^+, Na^+, K^+ and Cs^+ is 505, 410, 335 and 264 kJ/g-ion respectively [324].

As for the reason for poor channel permeability to anions, this could be large anion sizes ($r_{Cl^-} = 1.81$ Å, $r_{Br^-} = 1.93$ Å, $r_{I^-} = 2.2$ at $r_{H_2O} = 1.85$ Å) and the significant heat value of anion hydration.

However, despite their large size, the fraction of anions penetrating into the channels could be greater than that of cations. Amide groups located on a surface of selective layer interact with anions, as well as amides on PVP or PVCL chains in aqueous solutions [197]. The constant of binding of PVP and anions grows in a series Cl^-, Br^- and I^- (see Chapter 4) because of their interaction with water molecules polarized by PVP.

Therefore anions could penetrate the channels prior to cations because of the interaction with water molecules polarized by dipoles of polymer material (amide, carboxylic, ester or ether groups). In fact, in a series Cl^-, Br^- and I^- water transfer selectivity at RO process is considerably reduced [328].

High local concentration of amide (ether or ester) groups on the walls of transportation ways (channels) is the key factor defining migration of water molecules inside the channel. Carbonyls act as bridges in 'hopping' transport of water molecules inside the channel under high external pressure.

```
                    H
                    |
                   .O..
               H  ´    `H
                        |      O=
      =O.               O
         ´´.    H     .  \
      NH      \  .O´´    ´O /
           H  ´    `O´    `H
              H        \      .
      =O         .O´´    H   O=
         ´.    .´   \        `.
        H   `O´  `H   H     \
             H        H      O ···HN
                              /
                            H .
                               `O=
```

Amide dipoles, surrounding and polarizing water molecules inside the channel, weaken interaction of the H_2O molecules, both between each other and with the matrix. This statement is made on the basis of experimental data and theoretical calculations on the hydration energy of the amide PVP group.

The fact is established of the frequency increase of stretching vibrations of the H—O bond in water molecule with the reduction of the number of water molecules (N) per PVP chainlink [120], that indicates a decrease of hydrogen bond energy at going from H—O···H—O to H—O···O=C— bonds. Quantum-chemical calculations showed that the heat of hydrogen bond formation between C=O (pyrrolidone ring) and the first water molecule is less than those of hydrogen bonds between the subsequent water molecules (see Chapter 3). The significant decrease of melting heat of ice-like formations in poly-*N*-vinylamide solutions and in PVCL gels with the increase of polymer concentration clearly demonstrates the effect of dipoles on hydrogen bond rupture between water molecules, accompanied by growth of the number of molecules with incomplete realization of hydrogen bonds.

It should be noted that in PVCL gels at a small number of water molecules per chainlink (N = 3–6) the areas (channels) limited by rings are probably formed, where several water molecules are subjected to the action of C=O dipoles, as occurs in channels of polyamide RO membranes. Melting heat of ice-like formations in this surrounding decrease five to seven times in comparison with that of pure ice, indirectly indicating a significant structural transformation of water associates inside the gel and weakening of their interaction with a matrix.

In the review [349] the assessment of the self-diffusion coefficient of water molecules (D_{sd}) in the active layer was performed on the basis of salt and water rejection values, water and salt concentration inside a polymer selective layer and water flow at the RO process (25 °C, 1% NaCl, 96.7 atm pressure) on an example of aromatic (MPD + IPC) polyamide. The result of 1.44×10^{-6} cm^2/s was obtained.

Water mobility measurement by NMR with pulsing gradient magnetic field method [326] in swollen films of aromatic polyamides where the contents of fragments with sulfonate groups varied, showed that the D_{sd} value of water in these films is near to the D_{sd} specified above.

It turns out that the introduction of hydrophilic fragments in polyamide (CPA-1) chain obtained from MPD + 4,4′-diaminodiphenyl-2-sulfonic acid + IPC permits the organization of channels with certain contents of water molecules [353–356], simulating that of channels of a barrier layer in RO membranes made from aromatic polyamides. Therefore, in a film of pure poly-*m*-phenyleneisophthalamide with one or two water molecules per chainlink, the D_{sd} value is low and is 1.3×10^{-7} cm^2/s. At the increase of molar fraction (β) of amide fragments with a sulfonic group in Na$^+$-form in copolymer, D_{sd} value greatly increase and at $\beta = 36, 56$ and 69 mol% becomes $7.5 \times 10^{-7}, 1.4 \times 10^{-6}$ and 1.8×10^{-6} cm^2/s respectively. Channels of rigid design with four or five water molecules per chainlink, being stable in solutions of various salt concentration and having an Na$^+$ transport number close to unity are disposed only in a swollen film of polyamide of $\beta = 45$–56 mol% [359].

Hence, the mobility of water molecules in these narrow channels with amide groups on walls ($r = 4.5$–5.0) is rather high, being 20 times less than that in bulk water at 25 °C (2.8×10^{-6} cm^2/s).

Thus, poly-*N*-vinylamides of high local concentration of amide groups appear to be valuable model compounds providing important information on water properties in amide group surroundings that can be used not only for interpretation of physicochemical phenomena arising in synthetic polymers but also in finding specific properties of natural polymers with amide bonds and their behaviour in aqueous solutions and in an organism.

References

1. Reppe, W. (1949) *Acetylene Chemistry*. Ch A. Meyer, New York, p. 68.
2. Sidelkovskya, F. P. (1970) *Chemistry of N-vinylpyrrolidone and its Polymers*. Nauka, Moscow, p. 172 (in Russian).
3. Kononov, N. F., Ostrovsky, S. A. and Ustyniuk, L. A. (1977) *New Technology of Some Syntheses on the Basis of Acetylene*. Nauka, Moscow, p. 174 (in Russian).
4. Puetzer, B., Katz, L. and Horwitz, L. (1952) *J. Am. Chem. Soc.* **74**: 4959–4960.
5. Ushakov, S. N., Davidenkova, V. V. and Luschik, V. B. (1961) *Izv. AN SSSR OKhN*, pp. 901–906 (in Russian).
6. US Pat. (1954) 2 669 570.
7. Spanish Pat. (1952) 203 923.
8. US Pat. (1959) 2 891 058.
9. Vaculik, P. (1956) *Chemie monomeru*. Nakladatelstvi, Ceskoslovenske Akademie Ved, Praha (in Czech).
10. Shostakovsky, M. F., Kononov, N. F., Sildelkovskaya, F. P., Chekulaeva, I. A. and Zarutzky, V. V. (1968) in: *Chemistry of Acetylene*, ed. Shostakovsky, M. F., Nauka, Moscow, pp. 480–484 (in Russian).
11. Reppe, W., Herrle, K. and Fikentscher, H. Ger. Pat. (1958) 922 378 (in German).
12. Houvink, R. and Staverman, A. (1962) *Chemie und Technologie der Kunststoffe*. Akademische Verlagsgesellschaft, Leipzig (in German).
13. Kirsh, Yu. E., Karaputadze, T. M., Shumsky, V. I. and Bairamov, Yu. Yu. (1980) *Khim-Farm. Zh.* **1**: 79–80 (in Russian).
14. Kroker, R., Mueller, G. and Hofmann, E. (1991) US Pat. 5,039,817.
15. Veigang, K. and Hilgetag, G. (1964) *Organisch-Chemische Experimentierkunst*. Johan Ambrosis Barth, Leipzig (in German).
16. Odian, G. (1970) *Principles of Polymerization*. McGraw-Hill, New York.
17. Belgian Pat. (1963) 634 033.
18. Bestian, H. and Jensen, H. (1962) Ger. Pat. 1 176 124 (in German).
19. Bestian, H. and Hartwimmer, R. (1966) Ger. Pat. 1 196 657 (in German).
20. Kirsh, Yu. E. (1993) *Polymer Science* **35**: 271–285 (*Vysokomolek Soed*. **35B**: 98–114).
21. Blauche, R. B., Langmeadow, E. and Cohen, S. (1967) US Pat. 3 317 603.
22. Eck, H., Heckmaier, J. and Spes, H. (1976) Ger. Pat. 1 932 709 (in German).
23. Aksenov, A. I., Kondrat'ev, N. B., Demidova, N. V., Prokof'ev, E. P., Ovsepian, A. M., Lavut, E. A., Panov, V. P. and Kirsh, Yu. E. (1987) *Zh. Obshch. Khim.* **57**: 1634–1640 (in Russian).
24. Bestian, H. and Schnabel, H. (1970) Ger. Pat. 1 670 742 (in German).
25. Shnabel, H. (1973) Ger. Pat. 2 336 977 (in German).

26. Dawson, D. and Olleson, K. M. (1985) US Pat. 4 578 515.
27. Jpn Pat. (1986) 60-14955 (in Japanese).
28. Jpn Pat. (1988) 62-59248 (in Japanese).
29. Jpn Pat. (1988) 61-97309 (in Japanese).
30. Kroener, M., Schmidt, W. and Offring, A. (1985) Ger. Pat. 3 603 450 (in German).
31. Jpn Pat (1989) 63-190862 (in Japanese).
32. Linhart, F., Degen, H., Auhorn, W., Kroener, M., Hartmann, H., Heide, W (1988) US Pat. 4 772 359.
33. Pfoht, S., Kroener, M., Hartmann, H., Denzinger, W. (1988) US Pat. 4 774 285.
34. Brunnmuller, F., Kroener, M., Gotze, W. and Schmidt, W. (1984) Ger. Pat. 3 443 463 (in German).
35. Summerville, R. and Stackman, R. W. (1983) *Polymer Preprint* **24**: 12–15.
36. Dawson, D. J., Gless, R. D. and Wingard, R. E. (1978) *J. Am. Chem. Soc.* **98**: 5996–6001.
37. Stackman, R. W. and Summerville, R. H. (1985) *Ind. Eng. Chem. Prod. Res. Dev.* **24**: 242–247.
38. Ger. Pat. (1976) 2 503 114.
39. Akashi, M., Yashima, E., Yamashita, T., Miyanchi, N., Susita, S., and Marumo, K. (1990) *J. Pol. Sci., Pol. Chem.* **28**: 3487–3497.
40. Jensen, H., Schmidt, E. and Mitralaff, H. (1980) Ger. Pat. 2 919 755 (in German).
41. Schwiersch, W. and Hartwinner, R. (1967) Ger. Pat. 1 235 893 (in German).
42. Lynn, I. W. and Ash, B. D. (1969) US Pat. 3 144 396.
43. Kirsh, Yu. E. and Sokolova, L. V. (1983) *Khim-Farm. Zh.* **6**: 711–721 (in Russian).
44. Jpn Pat. (1987) 62-265304(A) (in Japanese).
45. Jpn Pat. (1988) 63-105009(A) (in Japanese).
46. Bovey, F. A. (1972) *High Resolution NMR of Macromolecules*. Academic Press, New York.
47. Kirsh, Yu. E., Berestova S. S., Aksenov A. I. and Karaputadze T. M. (1990) *Zh. Fiz. Khim.* **64**: 1894–1899 (in Russian).
48. Kirsh, Yu. E., Kalninsh, K. K., Pestov, D.V., Shatalov, G. V., Kuznetsov, V. A. and Krylov, A. V. (1996) *Russian J. Phys. Chem.* **70**: 802–806.
49. La Planche, L. A. and Rogers, M. T. (1964) *J. Am. Chem. Soc.* **86**: 337–344
50. Levy, G. C. and Nelson, G. L. (1973) *J. Am. Chem. Soc.* **94**: 4897–4902.
51. Levy, G. and Nelson, G. (1972) *Carbon-13 Nuclear Magnetic Resonance for Organic Chemists*. Academic Press, New York.
52. Robin, M. B., Bovey, F. A. and Basch, H. (1970) in: *The Chemistry of Amides*, ed. Zabicky J., Interscience, London.
53. Sewart, W. E. and Siddall, T. H. (1970) *Chem. Rev.* **70**: 517–551.
54. Kitano, M. and Kuchitsu, K. (1974) *Bull. Chem. Soc. Japan* **47**: 631–635.
55. Riddick, J. A. and Bunger, W. B. (1971) *Organic Solvents. Techniques of Chemistry*. Wiley-Interscience, New York.
56. Eliel, E. L., Allinger, N. L., Angyal, S. J. and Morrison, G. A. (1965) *Conformational Analysis*. Interscience, New York, London, Sydney.
57. Winkler, F. K. and Dunitz, J. D. (1975) *Acta Cryst.* **31**: 268–270.
58. Tischenko, G. N., Zhukhlistova, N. E. and Kirsh, Yu. E. (1997) *Crystallography Reports* **42**: 626–630.
59. Kirsh, Yu. E. Kokorin, A. I., Karaputadze, T. M. and Kazarin, L. I. (1981) *Vysokomol Soed.* **23B**: 444–447 (in Russian).
60. Rothschild, W. S. (1972) *J. Am. Chem. Soc.* **94**: 8676–8683.
61. Eley, D. D., Hey, M. I. and Winterngham, B. L. (1977) *J. Solution Chem.* **5**: 787–797.
62. Alimov, A. Kh., Karaputadze, T. M., Rashidova, S. Sh. and Kirsh, Yu. E. (1981) *Uzb. Khim. Zh.* **5**: 29–33 (in Russian).

63. Karaputadze, T. M., Bairamov, Yu. Yu., Shumsky, V. I., Kobiakov, V. V., Ovsepian, A. M., Panov, V. P. and Kirsh, Yu. E. (1982) *Proc. of II All-Union Conference 'Water-soluble Polymers and their Application'*. Irkutsk, Russia, p. 93 (in Russian).
64. Denisov, V. M., Ushakova, V. N., Koltzov, A. I. and Panarin, E. F. (1989) *Zh. Prikl. Khim.* **62**: 660–663 (in Russian).
65. Nowlan, V. J. and Tidweel, T. T. (1977) *Accounts Chem. Res.* **10**: 252–258.
66. Izvolensky, V. V., Drovenkova, N. V., Semchikov Yu. D., Sveshnikova, T. G. and Shalin, S. K. (1992) *Vysokomol. Soed.* **34A**: 53–59 (in Russian).
67. Fikentscher, H. and Herrle, K. (1945) *Modern Plastics* **23**: 157–163.
68. Schuster, C., Sauerbier, K., and Fikentscher, H. (1943) US Pat. 2 335 454.
69. Stein S (1958) in: *Chemiche Reactionen Ionisierender Strahlen (Radiation Chemistry)*. Herausgegeben von Mohler, H., Verlag, H. R. Saurlander Frankfurt am Main (in German).
70. Wessel, W., Schoog, M. and Winkler, E. (1971) *Arzneim-Forsch (Drug Res)* **21**: 1468–1482.
71. Herrle, K., Denzinger, W. and Seelert, K. (1974) Ger Pat. 2 439 196 (in German).
72. Herrle, K., Seelert, K., Denzinger, W. and Schwarz, W. (1977) US Pat. 4 027 083.
73. Herrle, K., Gansepohl, H. and Schwarz, W. (1977) US Pat. 4 053 696.
74. Kirsh, Yu. E., Karaputadze, T. M. *et al.* (1977). Inventor's Certificate of USSR 582 247. BI 30 (in Russian).
75. Kirsh, Yu. E., Karaputadze, T. M., Kochergin, P. M., Bairamov, Yu. Yu. and Shumsky, VI (1980) Inventor's Certificate of USSR 755 800 BI 30 (in Russian).
76. Kirsh, Yu. E., Ermolaev, A. V. and Karaputadze, T. M. (1981) *Khim-Farm. Zh.* **12**: 56–63 (in Russian).
77. Nuber, A., Lang, S., Sanner, A. and Schroder, G. (1987) Ger. Pat. 3 532 747 A1 (in German).
78. Shtamm, E. V., Karaputadze, T. M., Kirsh, Yu. E., Purmal', A. P. and Skurlatov, Yu. I. (1981) *Zh. Fiz. Khim.* **55**: 2289–2294 (in Russian).
79. Kozlov, Yu. N., Nadezhina, A. L. and Purmal', A. P. (1974) *Int. J. Chem. Kinetics* **6**: 383–396.
80. Davis, J. K. and Senoglas, E. (1981) *Austr. J. Chem.* **34**: 1413–1421.
81. Karaputadze, T. M., Shumsky, V. N., Skurlatov, Yu. I. and Kirsh, Yu. E. (1982) *Vysokomol. Soed.* **24B**: 305–309 (in Russian).
82. Gulis, I. M., Evdokimenko, V. M., Kirsh, Yu. E. and Lapkovsky, M. A. (1991) *Khim-Farm. Zh.* **9**: 82–86. (in Russian).
83. GB Pat. 1 513 258 (1975).
84. Davison, A. and Saugster, D. F. (1976) *J. Polym. Sci., Symp.* **55**: 249–257.
85. Shostakovsky, M. F., Sidelkovskaya, F. P. and Homutov, A. M. (1955) *Izv. AN SSSR OKhN.* **5**: 919–924 (in Russian).
86. Eisele, M. and Burchard, M. (1990) *Macromolek. Chem.* **191**: 169–184.
87. Solomon, O. F., Corciovei, M., Ciuta, I. and Boghina, C. (1968) *J. Appl. Pol. Sci.* **12**: 1835–1842.
88. Fisher, J. P. and Rosinger, S. (1983) *Macromol. Chem.* **184**: 1247–1253.
89. Lai, T. W. and Vijayendran, B. R. (1989) US Pat. 4 795 770
90. Badesso, R. J., Lai, T. W., Pinschmidt, R. K., Sagl, D. J. and Vijayendran, B. R. (1991) *Polym. Prep., Am. Chem. Soc., Div. Polym. Chem.* **32**: 110–114.
91. Bagdasarian, H. S. (1966) *Theory of Radical Polymerization*. Nauka, Moscow, p. 300 (in Russian).
92. House, D. A. (1962) *Chem. Rev.* **62**: 185–203.
93. Ferguson, J. and Rajan, V. S. (1979) *Eur. Polymer. J.* **15**: 627–631.
94. Panarin, E. F., Uzhakova, V. N., Lelinkh, A. I., Kirukhin, D. P. and Munihes, V. M. (1994) *Radiat. Phys. Chem.* **43**: 509–513

95. Cobianu, N., Marculesku, B., Bognina, C., Vasilescu, S. D. and Matache, S. (1973) *Materiale Plastice* **10**: 75–78.
96. Kirsh, Yu. E., Karaputadze, T. M., Shumsky, V. I. *et al.* (1990) Inventor's Certificate of USSR 1 613 446 BI (1990) **46**: 96.
97. Karaputadze, T. M., Shumsky, V. I. and Kirsh, Yu. E. (1978) *Vysokomol. Soed.* **20A**: 1854–1858 (in Russian).
98. Senoglas, E. and Thomas, R. A. (1978) *J. Polym. Sci., Pol. Lett.* **16**: 555–562.
99. Gordon, A. J. and Ford, R. A. (1972) *The Chemist's Companion*. Wiley, New York.
100. Shtamm, E. V., Scurlatov, Yu. I., Karaputadze, T. M., Kirsh, Yu. E. and Purmal', A. P. (1980) *Vysokomol. Soed.* **22B**: 420–423 (in Russian).
101. Kaplan, R. H. and Rodrigues, F. (1975) *J. Appl. Polym. Sci. Pol. Symp.* **26**: 181–186.
102. Agasandian, V. A., Grossman, E. A., Bagdasarian, H. S., Litmanovich, A. D. and Stern, V. A. (1966) *Vysokomol. Soed.* **8**: 1580–1585.
103. Cheng, H. M., Smith, T. E. and Vitus, D. M. (1981) *J. Polym. Sci. Pol. Let.* **19**: 29–31.
104. Ebdon, J. R., Huckerby, T. N. and Senogles, E. (1983) *Polymer* **2**: 697–703.
105. Yakimov, S. A., Kirsh, Yu. E. and Sibeldina, L. A. (1986) *Zh. Fiz. Khim.* **60**: 1295–1297.
106. Kirsh, Yu. E. (1993) *Prog. Polym. Sci.* **18**: 519–542.
107. Levy, G. C. and Nelson, G. L. (1964) *J. Amer. Chem. Soc.* **86**: 337–343.
108. Kirsh, Yu. E. Ermolaev, A. V., Karaputadze, T. M. *et al.* (1980) Inventor's Certificate of USSR 907 008 BI (1982) 7 (in Russian).
109. Englinskaya, L. V., Sheluhina, G. D., Letunova, A. B., Shaps, N. A., Karaputadze, T. M., Ermolaev, A. V. and Kirsh, Yu. E. (1982) *Khim-Farm. Zh.* **4**: 74–78 (in Russian).
110. Karaputadze, T. M., Shumsky, V. I., Sheluhina, G. D. and Kirsh, Yu. E. (1985) *Khim-Farm. Zh.* **3**: 212–216 (in Russian).
111. Robinson, B. V., Sullivan, F. M., Borzelleca, J. F. and Schwartz, S. L. (1990) *PVP. A Critical Review of the Kinetics and Toxicology of Polyvinylpyrrolidone (Povidone)*. Lewis, Michigan, p. 209.
112. Staszewska, U. (1983) *Angew Macromol Chem* **118**: 1–17.
113. Von Hengstenberg, J. and Schuch, E. (1951) *Macromolek. Chem.* **7**: 236–258.
114. Bender, M. (1964) *Catalytic Mechanism of Nucleophilic Reactions of Carboxylic Acid Derivatives*. Mir, Moscow, p. 192 (in Russian).
115. Denzinger, W. and Herrle, K. (1976) Ger. Pat. 2 514 127.
116. Karaputadze, T. M., Svergun, V. I., Tarabakin, S. V., Panov, V. P. and Kirsh, Yu. E. (1979) *Khim-Farm. Zh.* **10**: 119.
117. Grosser, F. (1960) US Pat. 2 938 017.
118. Ger. Pat. 1 268 391 (1960).
119. Hoffmann, E. and Herrle, K. (1972) Ger. Pat. 2 059 484.
120. Kobiakov, V. V., Ovsepian, A. M. and Panov, V. P. (1981) *Vysokomol. Soed.* **28A**: 150–160 (in Russian).
121. Pashkin I. I., Kirsh Yu. E., Zubov, V. P., Anisimova, T. V., Kuzkina, I. F., Voloshina, Ya. V. and Krylov, A. V. (1993) *Vysokomol. Soed.* **35A**: 481–484 (in Russian).
122. Bellamy L. J. (1957) *The Infra-Red Spectra of Complex Molecules*. Wiley, New York, Methuen, London.
123. Topchiev, D. A., Martynenko, A. I., Kabanova, E. Yu., Oppengeim, V. D., Kirsh, Yu. E. and Karaputadze, T. M. (1990) *Izv. AN SSSR ser Khim* **9**: 1969–1974 (in Russian).
124. Tophchiev, D. A., Martinenko, A. I., Kabanova, E. Yu., Timofeeva, L. M., Oppengeim, V. D., Shashkov, A. S. and Driabkina, A. I. (1994) *Vysokomol. Soed.* **36A**: 1242–1253 (in Russian).
125. Kirsh, Yu. E., Semina, N. V., Kalninsh, K. K. and Shatalov, G. V. (1996) *Polymer Science* **38B**: 423–426.

126. Fedorov, E. K., Lobanov, O. E., Mosalova, L. F., Svergun, V. I., Kedik, S. A. and Kirsh, Yu. E. (1994) *Polymer Science* **36A**: 1200–1204.
127. Kachahmadze, Z. N., Ovsepian, A. M., Karaputadze, T. M., Papunidze, G. R., Lazishvili, L. A., Panov, V. P. and Kirsh, Yu. E. (1988) *Vysokomol. Soed.* **31B**: 684–686 (in Russian).
128. Izvolensky, V. V., Semchikov, Yu. D. and Sveshnikova, T. G. (1995) *Proceedings of IV All-Union Conference 'Water-soluble Polymers and their Application'*, Irkutsk, Russia, p. 25.
129. Izvolenskii, V. V., Semchikov, Yu. D., Sveshnikova, T. G. and Shalin, S. K. (1992) *Vysokomol. Soed.* **34A**: 53–59 (in Russian).
130. Shapiro, A. and Le Doan Trung (1974) *Eur. Pol. J.* **10**: 1103–1106.
131. Nadzhimutdinov, Sh., Turaev, A. S., Usmanov, H. U., Usmanov, A. H. and Chulpanov, K. (1976) *Dokl. AN SSSR* **226**: 1113–1116 (in Russian).
132. Solovsky, M. V., Ushakova, V. N., Panarin, E. F., Boimirzaev, A. S., Nesterov, V. V., Peranen, A. A. and Mihalchenko, G. A. (1987) *Khim. Vysok. Energ.* **21**: 143–147 (in Russian).
133. Bork, J. and Coleman, L. E. (1960) *J. Polym. Sci.* **43**: 413–421.
134. Kahn, D. J. and Horowitz, H. H. (1961) *J. Polym. Sci.* **54**: 363–374.
135. Efremov, T. B., Meos, A. I., Volf, L. A. and Zarutskii, V. V. (1969) *Zh. Prikl. Khim.* **42**: 1196–1201 (in Russian).
136. Semchikov, Yu. D., Riabov, A. V. and Kashaeva, V. N. (1970) *Vysokomol. Soed.* **12B**: 381–384 (in Russian).
137. Narito, H., Hoshii, J. and Mashid, S. (1976) *Angew. Macromolek. Chem.* **52B**: 117–125.
138. Peppas, N. A. and Gehr, T. W. B. (1979) *J. Appl. Polym. Sci.* **24**: 2159–2167.
139. Sabey, M. Z., Dmitrieva, S. N. and Meos, A. N. (1970) *Vysokomol. Soed.* **12B**: 243–246.
140. Skorikova, E. E., Karaputadze, T. M., Ovsepian, A. M., Aksenov, A. P. and Kirsh, Yu. E. (1985) *Vysokomol. Soed.* **27B**: 869–872 (in Russian).
141. Gruz, R. N., Shibanovich, V. G. and Paparin, E. F. (1968) *Vysokomol. Soed.* **10A**: 2096–2101 (in Russian).
142. Panarin, E. F., Gavrilova, N. N. and Nesterov, V. V. (1978) *Vysokomol. Soed.* **20B**: 66–69.
143. Kopeikin, V. V. and Panarin, E. F. (1977) *Vysokomol. Soed.* **19A**: 861–866. (in Russian).
144. Panarin, E. F. and Gavrilova, N. N. (1977) *Vysokomol. Soed.* **19B**: 251–254 (in Russian).
145. Kirichenko, D. V., Izvolenskii, V. V. and Semchikov, Yu.D. (1995) *Vysokomol. Soed.* **37B**: 1953–1956 (in Russian).
146. Kathman, E. E. L. and McCormick, C. L. (1993) *Macromolecules* **26**: 5249–5252.
147. Scholtan, W. (1951) *Macromolek. Chem.* **7**: 209–235
148. Pavlov, G. M., Panarin, E. F., Korneeva, E. V., Kurochkin, K. V., Baikov, V. E. and Ushakova, V. N. (1990) *Vysokomol. Soed.* **32A**: 1190–1195 (in Russian).
149. Miller, M. E. and Hamm, F. A. (1953) *J. Phys. Chem.* **57**: 110–122.
150. Frank, H. P. and Levy, G. B. (1953) *J. Polym. Sci.* **10**: 371–378.
151. Levy, G. B. and Frank, H. P. (1955) *J. Polym Sci.* **17**: 247–254.
152. Graham, W. D. (1957) *J. Pharm. Pharmacol.* **9**: 230–236.
153. Kirsh, Yu. E., Karaputadze, T. M., Svergun, V. N., Soos, T. A., Sutkevich, L. A., Tverdohlebova, N. P. and Panov, V. P. (1980) *Khim-Farm. Zh.* **7**: 107–110 (in Russian).
154. Buhler, V. and Klodwig, U. (1984) *Acta. Pharm. Techn.* **30**: 317–324.
155. Senak, L., Wu, C. S. and Malawer, E. G. (1987) *J. Liquid. Chrom.* **10**: 1127–1135.
156. Sheluhina, G. D., Karaputadze, T. M., Ostrovsky, S. A., Kasaikin, V. A. and Kirsh, Yu. E. (1982) *Proceedings of II All-Union Conference 'Water-soluble Polymers and their Application'*, Irkutsk, Russia, p. 74.

157. Vasiliev, P. S., Suzdaleva, V. V., Fedorov, N. A. and Grozdev, D. M. (1976) in: *Problems of Hematology and Transfusion.* Moscow (in Russian). vol. 1, pp. 141–151.
158. Reske-Neielsen, E., Bojsen-Moller, M., Ventner, M. and Hansen, J. C. (1976) *Acta Path. Microbiol. Scand.* **84**: 397–405.
159. Sanner, A., Straub, F. and Tschang, Ch. (1983) *Proceedings of the International Symposium on Povidone,* eds Digenis, G. A. and Ansell, J., University of Kentucky, USA.
160. Ravin, H. A., Seligman, A. M. and Fine, J. (1952) *New Engl. J. Med.* **247**: 921–925.
161. Hespe, W., Meier, A. M. and Blankwater, J.I. (1977) *Arzneim-Forsch* **27**: 1158–1162.
162. Fleig, G. E. and Rodriguez, F. (1982) *Chem. Eng. Commun.* **13**: 219–224.
163. Nogemi, H., Nagai, T. and Kondo A. (1982) *Chem. Pharm. Bull.* **18**: 2280–2284.
164. Malawer, E. G., De Vasto, J. K., Frankoski, S. P. and Montana, A. J. (1984) *J. Liquid. Chrom.* **7**: 441–461.
165. Fedorov, E. K. and Kedik, S. A. (1994) *Polymer Science* **36B**: 1291–1294.
166. Kirsh, Yu. E., Soos, T. A., Karaputadze, T. M., Kobiakov, V. V., Sinitzina, L. A. and Ostrovsky, S. A. (1979) *Vysokomol. Soed.* **21A**: 2734–2740 (in Russian).
167. Kirsh, Yu. E., Soos, T. A. and Karaputadze, T. M. (1983) *Europ. Pol. J.* **19**: 639–645.
168. Sergeev, N. M. (1981) *NMR Spectroscopy.* MSU, Moscow, p. 76.
169. Yakimov, S. A., Shumsky, V. I., Kirsh, Yu. E., Sibeldina, L. A. and Karaputadze, T. M. (1986) *Zh. Fiz. Khim* **60**: 1291–1294.
170. Kirsh, Yu. E., Yanul', N. A., Karaputadze, T. M. and Timashev S. F. (1992) *Russian J. Phys. Chem.* **66**: 1399–1401.
171. Sinyukov, V. V. (1976) *Structure of Monoatomic Liquids, Water, and Aqueous Solutions of Electrolytes.* Nauka, Moscow, p. 256 (in Russian).
172. Gordon, J. E. (1975) *The Organic Chemistry of Electrolyte Solutions.* Wiley-Interscience, New York, London, Sydney, Toronto.
173. Ikada, Y., Suzuki, M. and Ivata, H. (1980) in: *Water in Polymers.* ed. Rowland, P., ACS Symposium Series 127, American Chemical Society, Washington, DC.
174. Danford, M. D. and Levy, H. A. (1962) *J. Am. Chem. Soc.* **84**: 3965–3966.
175. Samoilov, O. Ya. (1963) *Zh. Strukt. Khim.* **4**: 499–501 (in Russian).
176. Samoilov, O. Ya. and Nosova, T. A. (1965) *Zh. Strukt. Khim.* **6**: 798–808 (in Russian).
177. Frank, H. S. and Wen, W-Y. (1957) *Discuss. Faraday Soc.* **24**: 13–17.
178. Yebb, J. L. (1963) *Enzyme and Metabolic Inhibitors.* Academic Press, New York, London.
179. Luck, W. A. P. (1980) in: *Water in Polymers,* ed. Rowland, S. P., ACS Symposium Series 127, American Chemical Society Washington, DC.
180. *Structure and Stability of Biological Macromolecules,* eds Timasheff, S.N. and Fasman, G. G., Marcel Dekker, New York.
181. Kirsh, Yu. E., Yanul, N. A., Bakeeva, I. V. and Pashkin, I. I. (1996) *Proc of 2nd International Symposium 'Molecular Order and Mobility in Polymer Systems',* St Petersburg, p. P-027.
182. Kirsh, Yu. E., Yanul, N. A., Pashkin, I. I., Bakeeva, I. V., Zubov, V. P. and Timashev, S. F. (1998) *Zh. Fiz. Khim.* **72**: (to be published).
183. Quinn, F. X., Kampff, E., Smyth, G. and McBrierty, V. J. (1988) *Macromolecules* **21**: 3191–3198.
184. McBrierty, V. J., Quinn, F. X., Keely, C., Wilson, A. C. and Friends, G. D. (1992) *Macromolecules* **25**: 4281–4284.
185. Ahmad, M. B., Huglin, M. B. (1994) *Polymer International* **33**: 273–277.
186. Deodhar, S. and Luner, Ph. (1980) in: *Water in Polymers,* ed. Rouland, P., ACS Symposium Series 127, American Chemical Society, Washington, DC.
187. Wasserman, A. M. and Kovarsky, A. L. (1986) *Spin Labels and Probes in Physico-chemistry of Polymers.* Nauka, Moscow, p. 241 (in Russian).

188. Aleksandrova, T. A., Karaputadze, T. M., Shapiro, A. B., Kirsh, Yu. E. and Wasserman, A. M. (1982) *Vysokomol. Soed.* **24A**: 2373–2378 (in Russian).
189. Wasserman, A. M., Timofeev, V. P., Aleksandrova, T. A., Karaputadze, T. M., Shapiro, A. B. and Kirsh, Yu. E. (1983) *Europ. Pol. J.* **19**: 333–339.
190. Anufrieva, E. V., Ramazanova, M. R., Luschik, V. B., Nekrasova, T. N., Sheveleva, T. V., Karaputadze, T. M. and Kirsh, Yu. E. (1986) *Vysokomol. Soed.* **28A**: 573–576 (in Russian).
191. Kirsh, Yu. E. (1985) *Prog. Pol. Sci.* **11**: 283–338.
192. Wasserman, A. M., Aleksandrova, T. A. and Kirsh, Yu. E. (1980) *Vysokomol. Soed.* **22A**: 275–281 (in Russian).
193. Wasserman, A. M., Aleksandrova, T. A., Kirsh, Yu. E. and Buchachenko, A. L. (1979) *Europ. Pol. J.* **15**: 1051–1055.
194. Kirsh, Yu. E., Krylov, A. V., Belova, T. A., Abdel'sadek, C. G. and Pashkin, I. I. (1996) *Russian J. Phys. Chem.* **70**: 1302–1306.
195. Gordon, J. E. (1972) *J. Am. Chem. Soc.* **94**: 650–656.
196. Rabinovich, I. M. (1972). *Application of Polymers in Medicine*, Medicine, Leningrad p. 177 (in Russian).
197. Yakimov, S. A., Kirsh, Yu. E. and Sebildina, L. A. (1987) *Zh. Fiz. Chim.* **62**: 3350–3354.
198. Fisher, J. J. and Jardetzky, O. (1965) *J. Am. Chem. Soc.* **87**: 3237–3244.
199. Landay, M. A. (1981) *Molecular Mechanisms of Action of Physiologically Active Compounds (Moleculyarnye mekhanizmy deistviya fiziologicheski aktivnykh soedineni).* Nauka, Moscow (in Russian).
200. Kirsh, Yu. E., Yakimov, S. A., Sebildina, L. A. and Karaputadze, T. M. (1988) *Zh. Fiz. Khim.* **62**: 711–714 (in Russian).
201. Saito, S. and Yukawa, M. (1969) *Kolloid J.* **254**: 1015–1017.
202. Saito, S. and Yukawa, M. J. (1969) *J. Colloid Interface Sci.* **30**: 211–218.
203. Saito, S. (1970) *J. Polym. Sci.* **1A**: 263–271.
204. Shelanski, M.A. (1951) *Chem. Eng. News* **29**: 664–667.
205. Oster, G. and Immergut, E. N. (1955) *J. Am. Chem. Soc.* **76**: 1393–1396.
206. Neel, J. and Sebille, B. (1961) *J. Chim. Phys.* **58**: 738–753.
207. Palmer, D. A. and Mesmer, R. E. (1984) *J. Solution Chem.* **13**: 673–679.
208. Soos, T. A., Karaputadze, T. M., Bairamov, Yu. Yu., Kazarin, L. A. and Kirsh, Yu. E. (1981) *Vysokomol. Soed.* **23À**: 631–635 (in Russian).
209. Kirsh, Yu. E., Yakimov, S. A., Dolotova, T. A., Sibeldina, L. A. and Karaputadze, T. M. (1986) *Zh. Fiz. Khim.* **60**: 1027–1031 (in Russian).
210. Sillescu, H. and Brusau, R.G. (1970) *Chem. Phys. Lett.* **5**: 525–526.
211. Rockelmann, H. and Sillescu, H. (1974) *Z. Phys. Chem.* **92**: 263–280.
212. Yakimov, S. A., Kirsh, Yu. E., Sebildina, L. A. (1987) *Zh. Fiz. Khim.* **61**: 3347–3349.
213. Klotz, I. M., Walker, F. and Pivau, R. (1946) *J. Am. Chem. Soc.* **68**: 1486–1490.
214. Inoe, M. and Otsu, T. (1976) *J. Pol. Sci.* **14**: 1939–1944.
215. Klotz, I. M. and Shikawa, K. (1968) *Archives Biochem. Biophys.* **123**: 551–557.
216. Molyneux, P. and Frank, H. P. (1961) *J. Am. Chem. Soc.* **83**: 3169–3174 and 3175–3180.
217. Plaizier-Vercammen, J. C. and De Neve, R. E. (1981) *J. Pharm. Sci.* **71**: 552–556.
218. Scholtan, W. (1953) *Makromolek. Chem.* **11**: 131–230.
219. Takagishi, T. and Kuroki, N. (1973) *J. Pol. Sci. Pol. Chem. Ed.* **11**: 1889–1990.
220. Takagishi, T., Imajo, K., Nakagami, K. and Kuroki, N. (1977) *J. Pol. Sci. Pol. Chem. Ed.* **15**: 31–38.
221. Maruthamuthu, M. and Sobhana, M. (1979) *J. Pol. Sci. Pol. Chem. Ed.* **17**: 3159–3167.
222. Rodionova, N. A., Semisotnov, G. V., Kutyshenko, V. P., Uversky, V. N., Bolotina, I. A., Bychkova, V. E. and Ptitzin, O. B. (1989) *Molek. Biologiya.* **23**: 683–689 (in Russian).

223. Semisotnov, G. V., Rodionova, N. A., Razgulyaev, O. I., Uversky, V. N., Gripas, A. F. and Gilmanshin, R. I. (1991) *Biopolymers* **31**: 119–126.
224. Vladimirov, Yu. A. and Dobretzov, G. E. (1980) *Fluorescent Probes in Research of Biological Membranes*. Nauka, Moscow (in Russian).
225. Anufrieva, E. V., Nekrasova, T. N., Sheveleva, T. V. and Krakoviak, M. G. (1994) *Vysokomol. Soed.* **36A**: 449–456 (in Russian).
226. Kirsh, Yu. E. Soos, T. A., Kobiakov, V. V. and Panov, V. P. (1976) *Vysokomol. Soed.* **18A**: 388–389 (in Russian).
227. Kirsh, Yu. E., Soos, T. A. and Karaputadze, T. M. (1977) *Vysokomol. Soed.* **19A**: 2772–2779 (in Russian).
228. Takagishi, T., Naoi, J. and Kuroki, N. (1977) *J. Pol. Sci. Pol. Chem. Edit.* **15**: 2789–2785.
229. Gotlib, Yu. Ya., Pavlova, N. R., Kirsh, Yu. E. and Kabanov, V. A. (1977) *Vysokomol. Soed.* **19A**: 1150–1157 (in Russian).
230. Killman, E. and Bittler, R. (1972) *J. Pol. Sci. Part C* **39**: 247–263.
231. Valuev, L. I., Vakula, N. V. and Plate, N. A. (1984) *Vysokomol. Soed.* **24A**: 1700–1705 (in Russian).
232. Shopovalenko, E. P. and Kolosov, I. V. (1978) *Bioorganicheskaya Khimiya* **4**: 514–518 (in Russian).
233. Fishman, M. L. and Eirich, F. R. (1971) *J. Phys. Chem.* **75**: 3135–3140.
234. Fishman, M. L. and Eirich, F. R. (1975) *J. Phys. Chem.* **79**: 2740–2745.
235. Anufrieva E. V., Panarin, E. F., Pautov, V. D. and Solovsky, M. V. (1981) *Vysokomol. Soed.* **23A**: 1222–1228.
236. Pautov, V. D., Anufrieva, E. V., Kirpatch, A. B., Panarin, E. F., Gavrilova, I. I., Kochetkova, I. S., Luschik, V. B., Solovsky, M. V. and Ushakova, V. N. (1988) *Vysokomol. Soed.* **30A**: 2219–2224.
237. Pautov, V. D., Kirpatch, A. B., Anufrieva, E. V. and Panarin, E. F. (1990) *Vysokomol. Soed.* **32A**: 133–136.
238. Pautov, V. D., Kirpatch, A. B., Panarin, E. F., Gavrilova, I. I. and Kochetkova, I. S. (1987) *Proceedings of 3rd All-Union Conference 'Water-soluble Polymers and their Application'*. Irkutsk, Russia, p. 133.
239. Tenford, C. (1961) *Physical Chemistry of Macromolecules*. Wiley, New York.
240. Anufrieva, E. V., Nekrasova, T. N., Luschik, V. B., Fedotov, Yu. A., Kirsh, Yu. E. and Krakoviak, M. G. (1992) *Vysokomol. Soed.* **34B**: 31–34 (in Russian).
241. Kirsh, Yu. E., Fedotov, Yu. A., Iudina, N. A., Artemov, D. Yu., Yanul', N. A. and Nekrasova, T. N. (1991) *Vysokomol. Soed.* **33A**: 1127–1133 (in Russian).
242. Kabanov, V. A. and Papisov, I. M. (1979) *Vysokomol. Soed.* **21A**: 243–281 (in Russian).
243. Zezin, A. B. and Kabanov, V. A. (1982) *Uspekhi Khimii* **51**: 1447–1483 (in Russian).
244. Kirsh, Yu. E., Aleksandrova, T. A. and Wasserman, A. M. (1980) *Vysokomol. Soed.* **22**: 45–47 (in Russian).
245. Antipina, A. D., Papisov, I. M. and Kabanov, V. A. (1970) *Vysokomol. Soed.* **12A**: 329–331 (in Russian).
246. Anufrieva, E. V., Ramazanova, M. R., Krakoviak, M. G., Luschik, V. B., Nekrasova, T. N. and Sheveleva, T. V. (1991) *Vysokomol. Soed.* **33A**: 1186–1191 (in Russian).
247. Anufrieva, E. V., Ramazanova, M. R., Krakoviak, M. G., Luschik, V. B., Nekrasova, T. N. and Sheveleva, T. V. (1989) *Vysokomol. Soed.* **33A**: 256–261 (in Russian).
248. Anufrieva, E. V. and Gotlib, Yu. Yu. (1981) *Advances in Polymer Sci.* **40**: 1–57.
249. Anufrieva, E. V. and Pautov, V. D. (1992) *Vysokomol. Soed.* **34A**: 41–47 (in Russian).
250. Burget, Y. and Zundel, G. (1987) *Biopolymers* **86**: 95–108.
251. Burget, Y. and Zundel, G. (1987) *Biophys J.* **52**: 1065–1070.

252. Sherstyuk, S. F., Galaev, I. Yu., Savitsky, A. I., Kirsh, Yu. E. and Berezin, I. V. (1987) *Biotekhnologiya* **3**: 179–183 (in Russian).

253. Kuz'kina, E. F., Pashkin, I. I., Markvicheva, E. A., Kirsh, Yu. E., Bakeeva, I. V. and Zubov, V. P. (1996) *Khim-Farm. Zh.* **1**: 39–41 (in Russian).

254. Markvicheva, E. A., Kuz'kina, E. F., Pashkin, I. I., Plechko, T. N., Kirsh, Yu. E. and Zubov, V. P. (1991) *Biothechnol Techn.* **5**: 223–226.

255. Brandts, J. F. (1969) *Structure and Stability of Biological Macromolecules*, eds Timasheff, S. N. and Fasman, G. G., Marcel Dekker, New York.

256. *Immobilized Enzymes* (*Immobilizovannye Fermenty*) (1982) Eds Berezin, I. V., Antonov, V. K. and Martinek, K., MGU, Moscow, p. 75 (in Russian).

257. Kirsh, Yu. E., Galaev, I. Y. I., Karaputadze, T.M., Margolin, A. L., and Shvyadas, V. K. (1987) *Biotekhnologiya* **3**: 184–189 (in Russian).

258. Markvicheva, E. A., Bronin, A. S., Kudryavtseva, N. E., Kuz'kina, I. F., Pashkin, I. I. Kirsh, Yu. E. Rumsh, L. D. and Zubov, V. P. (1994) *Bioorganicheskaya Khimiya* **20**: 257–264 (in Russian).

259. Johynsson, A. C. and Mosbach, K. (1974) *Biochim. Biophys. Acta* **370**: 339–347.

260. Yasui, T. and Ichihara, Y. (1975) Jpn. Pat. 75-53583.

261. Koshelev, S. A., Davidenko, T. I., Kirsh, Yu. E, Pashkin, I. I. and Kuz'kina, I. F. (1994) *Applied Biochemistry and Microbiology* **30**: 285–290.

262. Markvicheva, E. A., Kuz'kina, I. F. *et al* (1991) Inventor's Certificate of USSR 1 721 088, BI(1992)11.

263. Markvicheva, E. A., Mareeva, T., Bronin, A. and Khaidukov, S. (1991), *Abstracts of 8th Conference of Young Scientists on Organic and Bioorganic Chemistry*, Riga, p. 212.

264. Galaev, I. Yu, and Mattiason, B. (1992) *Biotechnol. Techn.* **6**: 353–358.

265. Syzdaleva, V. V. and Vidavskaya, G. M. (1973) *Proceedings of the 2nd Republic Congress of Gematology* (*Materialy 2 Respyblicanskogo Congressa Gematologov*) Minsk, pp. 267–268 (in Russian).

266. Abramova, E. E. and Fromm, A. A. (1962) *Pediatrics* (*Pediatriya*) **4**: 35–39 (in Russian).

267. Fromm, A. A., Vasiliev, P. C. and Syzdaleva, V. V. (1966) *Problems of Gematology and Transfusion* (*Problemy Gematologii i Perilivaniya Krovi*) **7**: 17–19 (in Russian).

268. Fromm, A. A. and Sirotenko, A. V. (1964) *War-Medicine J.* (*Voenno-Meditsinsky Zh.*) **5**: 13–17 (in Russian).

269. Fromm, A. A. and Sirotenko, A. V. (1964) *Problemy Gematologii i Perilivaniya Krovi* **9**: 18–21 (in Russian).

270. Fromm, A. A. and Sirotenko, A. V. (1966) *Problemy Gematologii i Perlivaniya Krovi* **7**: 23–26 (in Russian).

271. Murazyan, R. I. and Strizhevskaya, L. N. (1972) *Proceedings of Scientific Works of R. O. Eloyana's Institute of Gematology and Transfusion* (*Tezisy Nauchnukh Rabot Instituta Gematologii i Perelivaniya Krovi imeni professora*, Eloyana R. O.) Erivan, pp. 75–76 (in Russian).

272. Kiselev, A. E., Rosenberg, A. E., Vasiliev, P. C., Grozdev, D. Ya., Agranenko, V. A. and Fromm, A. A. (1969) *Dictionary on Bloodsubstitutes on Preparations of Blood* (*Spravochnik pokrovesamenitelyam i preparatam krovi*) Moscow, pp. 3–41 (in Russian).

273. Kochin, N. N., (1974) *Proceedings of Kharkov Medicine Institute* (*Trudy Kharkovskogo Meditsinskogo Instityta*) **114**: 185–187 (in Russian).

274. Provotorov, V. M. and Nikitin, A. V. (1983) *Therapeutic Archives* (*Terapevtichesky Arkhiv*) **9**: 22–25 (in Russian).

275. Davatarova, M. M. (1983) *Bull. Dermatology and Venerealogy* (*Byulliten Dermatologii i Venerologii*) **3**: 41–46 (in Russian).

276. Ozinkovsky, V. V. (1979) *J. of Ear, Nose and Throat Diseases (Zh. Ushnykh. Nosovykhs. i Gorlovykh bolesney)* **4**: 97–98 (in Russian).
277. Kulaga, V. V., Litvinova, V. V., Latysheva, V. V. and Talovskaya, Zh. S. (1977) *Bull. Dermatology and Venerealogy (Byulliten Dermatologii i Venerologii)* **4**: 70 (in Russian).
278. Doshko, E. V. and Skygarevskaya, E. I. (1966) *Problems of Psychiatry (Voprosy psikhiatrii) Minsk* **3**: 87 (in Russian).
279. Kulemin, A. G. (1977) *Proceedings of Moscow Institute of Psychiatry (Trudy Mockovckogo Instityta Psikhiatrii MZ USSR)* **76**: 223–227 (in Russian).
280. Soroka, T. T. (1978) *Public Health of Belorussia (Zdravookhranenie Belorussia)* **5**: 47–49 (in Russian).
281. Chalisov MA and Korzhevsky EF (1974) *Proceedings of First congress of Neurophatology and Psychiatry (Materialy 1 kongressa nevrapotologov i psikhiatrov).* Belorussia, Minsk, pp. 266–268 (in Russian).
282. Krylova, O. M. and Smirnova, S. A. and Khlebnikova, I. M. (1968) *Therapeutic Archives (Terapevtich arkhiv)* **40**: 71–74 (in Russian).
283. Lapshina, A. V. (1973) *J. Ophthalmology (Vestnik oftalmalogii)* **2**: 57–59 (in Russian).
284. Semenov, A. D. (1973) *Materials of the 4th Ophthalmologic Congress of the USSR, Kiev* **1**: 524–525 (in Russian).
285. Zazhirei, V. D., Melnikova, E. P., Karaputadze, T. M., Chernenko, G. T. and Kirsh, Yu. E. (1985) *Khim-Farm. Zh.* **8**: 974–978 (in Russian).
286. Kirsh, Yu. E. and Zazhirei, V. D. (1984) in: *Reports of I All-Union Seminar on Synthetic Polymers of Medical Application,* eds. Plate, N. A. and Rashidova, S. Sh., FAN, Tashkent, pp. 154–162 (in Russian).
287. Kwitkowska, J. (1971) *Post. Hig. Med. Dosw.* **25**: 831–870.
288. Zimecki, M., Webb, D. R. and Rogers, Th. (1980) *J. Arch Immunol. Ther.* **24**: 179–197.
289. GB Pat. (1968) 1 131 007.
290. GB Pat. (1978) 1 508 601.
291. GB Pat. (1981) 1 592 053.
292. Berkengeim, T. B., Gyulbadamova, N. M. and Zhorov, V. I. (1971) *Experimental Surgery and Anesthesia* **6**: 70–72.
293. Zhorov, V. I., and Zhorov, I. S. (1974) *Sov Medetsina,* **10**: 49–51 (in Russian).
294. Zhorov, V. I., Zhorov, I. S. (1975) *Sov Medetsina* **7**: 62–64 (in Russian).
295. Zhorov, V. I., Kirsh, Yu. E., Shumsky, V. I., Karaputadze, T. M. and Bairamov, Yu. Yu. (1982) Inventor's Certificate of USSR 975 016. B (1983) 43.
296. Zhorov, V. I., Kirsh, Yu. E., Shumsky, V. I., Karaputadze, T. M. and Bairamov, Yu. Yu. (1983) US Pat. 4 389 404.
297. Zhorov, V. I. Kirsh, Yu. E.,Shumsky, V. I., Karaputadze, T. M. and Bairamov, Yu. Yu. (1983) GB Pat. 2 075 342.
298. Stanski, D. R., Paalzow, L. and Edlund, P. O. (1982) *J. Pharm. Soc.* **71**: 314–317.
299. Plate, N. A. and Vasiliev, A. E. (1986) *Physiologically Active Polymers.* Khimiya, Moscow, p. 294 (in Russian).
300. Solovsky, M. V., Pautov, V. D., Panarin, E. F., Kochetkova, N. S. and Afinogenov, G. E. (1983) *XX Scientific Conference, Institute of High Molecular Compounds,* RAN, Leningrad, pp. 84.
301. Panarin, E. F., Solovsky, M. V., Zaikina, N. A. and Afinogenov, G. E. (1985) *Makromol. Chem. Suppl.* **9**: 25–33.
302. Solovsky, M. V., Afinigenov, G. E., Panarin, E. F., Epanchinzeva, E. V. and Petukhova, N. A. (1991) *Khim-Farm. Zh.* **4**: 40–43.
303. Zaikina, N. A., Yagodkina, M. V. and Solovsky, M. V. (1994) *Antibiotics and Chemotherapy* **4**: 48–51 (in Russian).

304. Zaikina, N. A., Razin, A. N. and Solovsky, M. V. (1994) *Micology and Phytopathology* **6**: 32–34 (in Russian).
305. Afinigenov, G. E., Panarin, E. F. and Solovsky, M. V. (1977) in: *Preventive Maintenance and Treatment of Complex Infection Traumas.* Medicine, Leningrad, pp. 113–114 (in Russian).
306. Moravetz, H. (1965) *Macromolecules in Solution.* Interscience, New York, London, Sydney, p. 361
307. Frank, H. P. (1954) *J. Polymer Sci.* **12**: 565–573.
308. Conix, A. and Smets, G. (1955) *J. Polymer Sci.* **15**: 221–229.
309. Wolf, F., Lohs, K. and Bohm, S. (1970) *Macromolek. Chem.* **134**: 241–251.
310. Kirsh, Yu. E., Semina, N. V., Yanul', N. A. and Shatalov, G. V. (1994) *Zh. Fiz. Khim.* **68**: 1584–1586.
311. Anufrieva, E. V. (1982) *Pure and Appllied Chemistry* **54**: 533–548.
312. Krakoviak, M. G., Luschik, V. B., Ananieva, G. D., Panarin, E. F., Solovsky, M. V., Gorbunova, O. P., Gavrilova Ii., Kirsh, Yu. E., Pautov, V. D., Ramazanova, M. R. and Anufrieva, E. V. (1987)*Vysokomol Soed.* **29**: 598–603.
313. Krakoviak, M. G., Anufrieva, E. V., Luschick, V. B, Shelehov, N. S. and Skorodokhodov, S. S. (1978) *J. Macromolec. Sci., Chem.* **12**: 789–795.
314. Luschik, V. B., Krakovyak, M. G. and Skorodokhodov, S. S. (1980) *Vysokomol. Soed.* **22A**: 1904–1908.
315. Solovsky, M. V., Anufrieva, E. V., Panarin, E. F. Pautov, V. D. and Afinogenov, G. E. (1982) *Makromolek. Chem.* **183**: 1775–1780.
316. Kolodkina, I. I., Yurkevich, A. M., Panarin, E. F. (1976) *Vysokomol. Soed.* **18B**: 490–493.
317. Solovsky, M. V., Nazarova, O. V., Zubko, N. V. and Panarin, E. F. (1983) *Izv. AN USSR ser. Khim.* **3**: 685–687.
318. Solovsky, M. V. and Petukhova, N. A. (1992) *Vysokomol. Soed.* **34A**: 30–34.
319. Solovsky, M. V. (1994) *Zh. Prikl. Khim.* **68**: 1670–1674.
320. Solovsky, M. V., Panarin, E. F. and Gorbunova, O. P. (1995) *Zh. Prikl. Khim.* **69**: 1872–1876.
321. Solovsky, M. V., Denisov, V. M., Panarin, E. F., Petukhova, N. A. and Purkina, A. V. (1996) *Zh. Prikl. Khim.* **70**: 295–299.
322. Kirsh, Yu. E., Batrakova, M. V., Galaev, I. Yu., Aksenov, A. I. and Karaputadze, T. M. (1988) *Vysokomol. Soed.* **30A**: 365–369.
323. Poltorak, O. M. and Chukhrai, E. S. (1971) *Physico-chemical Principles of Enzymic Catalysis.* Vusshaya Shkola, Moscow, p. 311.
324. Nekrasov, B. V. (1962) *Obschaya Khimiya.* GKhI, Moscow, p. 973.
325. Karaputadze, T. M., Shumsky, V. I., Sheluhkina, G. D., Kirsanov, A. T. and Kirsh, Yu. E. (1983) *Khim-Farm. Zh.* **10**: 1251–1254 (in Russian).
326. Volkov, B. I., Korochkova, S. A., Nesterov, N. A., Fedotov, Yu. A., Kirsh, Yu. E. and Timashev, S. F. (1994) *Zh. Fiz. Khim.* **68**: 1310–1316 (in Russian).
327. Gritskova, I. A., Gromakova, I. V. and Kirsh, Yu. E. *Colloid Polym. Sci.* **274**: 884–888.
328. *Technological Processes with Application of Membranes* (1976) Mir, Moscow, p. 370 (in Russian).
329. Kesting, R. E. (1985) *Materials Science of Synthetic Membranes*, ed. Lloyd, D. R., ACS Symposium, series 269, ACS, Washington, DC, p. 131.
330. Nguyen, Q. T., Le Blanc, L. and Neel, J. (1985) *J. Membrane Sci.* **22**: 245–255.
331. Tam, C. M., Tweddle, T. A., Kutowy, O. and Hazlett, J. D. (1993) *Desalination* **89**: 275–287.
332. Guiver, M. D., Tam, C. M., Dal-Cin, M. M., Tweddle, T. A. and Kumar, A. (1995) *Proceedings of EUROMEMBRANE'95*, University of Bath, 1:1–345

333. Dal-Cin, M. M., Tam, C. M., Guiver, M. D. and Tweddle, T. A. (1994) *J. Appl. Polym. Sci.* **54**: 783–792

334. Miyano, T., Matsuura, T. and Sourirajan, S. (1993) *Chemical Engineering Comm.* **119**: 23–39.

335. Boom, R. M., van der Boomgaard, Th. and Smolders, C. A. (1994) *J. Membrane Sci.* **90**: 231–249.

336. Lhommeau, C. and Pith, T. (1995) *Proceedings of EUROMEMBRANE '95*, University of Bath, **1**: 1–349

337. Boom, R. M., Wienk, I. M., van der Boomgaard, Th. and Smolders, C. A. (1992) *J. Membrane Sci.* **73**: 272–292.

338. Strathman, H. (1985) *Material Science of Synthetic Membranes*, ed. Lloyd, D. R., ACS Symposium, Series 269, ACS, Washington, DC, p. 165.

339. Cabasso, I. (1985) *Material Science of Synthetic Membranes*, ed. Lloyd, D. R., ACS Symposium, Series 269, ACS, Washington, DC, p. 305.

340. Boom, R. M., Reinders, H. W., Rolevink, H. H. W., van der Boomgaard, Th. and Smolders, C. A. (1994) *Macromolecules*, **27**: 2041–2044.

341. Lonsdale, H. K. (1976) in: *Technological Processes with Application of Membranes*. Mir, Moscow, p. 133 (in Russian).

342. Lloyd, D. R. and Meluch, T. B. (1985) *Materials Science of Synthetic Membranes*, ed. Lloyd, D. R., ACS Symposium, Series 269, ACS, Washington, DC, p. 47.

343. Kirsh, Yu. E. and Timashev, S. F. (1991) *Zh. Fiz. Khim.* **65**: 2469–2478 (in Russian).

344. Timashev, S. F. (1988) *Physico-chemistry of Membrane Processes*. Khimiya, Moscow, p. 237 (in Russian).

345. Sourirajan, S., (1970) *Reverse Osmosis*. Academic Press, New York, p. 120.

346. Cadotte, J. E. (1985) *Materials Science of Synthetic Membranes*, ed. Lloyd, D. R. ACS Symposium, Series 269, Washington, DC, p. 273.

347. Hoehn, H. H. (1985) *Materials Science of Synthetic Membranes*, ed. Lloyd, D. R., ACS Symposium, Series 269, Washington, DC, p. 81.

348. Kirsh, Yu. E. and Popkov, Yu. M. (1988) *Uspekhi Khim.* **57**: 1001–1008.

349. Koros, W. J., Fleming, G. K., Jordan, S. M., Kim, T. H. and Hoehn, H. H. (1988) *Prog. Polym. Sci.* **13**: 339–401.

350. Kirsh, Yu. E. (1994) *Russian J. Appl. Chem.* **67**: 159–174.

351. Koo, J-Y., Petersen, R. J. and Cadotte, J. E. (1987) *Proceedings of the 1987 International Congress on Membranes and Membrane Processes (ICOM'87)*. Tokyo, Japan, p. 350.

352. Chen, J. Y., Kurihara, M. and Pusch, W. (1983) *Desalination* **46**: 379–388.

353. Kirsh, Yu. E. Malkina, I. M., Fedotov, Yu. A., Yanul', N. A., Gitis, S. S., Smirnov, S. A. and Timashev, S. F. (1993) *Vysokomol. Soed.* **35A**: 320–326 (in Russian).

354. Valvuev, V. V., Zemlianova, O. Yu., Semina, N. V., Fedotov, Yu. A. Kirsh, Yu. E. and Timashev, S. F. (1994) *Zh. Fiz. Khim.* **68**: 1681–1687 (in Russian).

355. Kirsh, Yu. E., Semina, N. V., Yanul', N. A., Malkina, I. M., Fedotov, Yu. A. and Timashev, S. F. (1995) *Elektrokhimiia* **35**: 11-17 (in Russian).

356. Kirsh, Yu. E. (1995) *Proceedings of EUROMEMBRANE '95*, University of Bath, **1**: 1–181.

357. Kurihara, M., Himeshima, Y. and Uemura, T. (1987) *ICOM '87*, Tokyo, p. 428.

358. Gleter, J. and Zachariah, M. R. (1985) *Reverse Osmosis and Ultrafiltration*, eds Sourirajan, S. and Matsuura, T., ACS Symposium Series 281, Washington, DC, p. 345.

359. Kirsh, Yu. E. (1993) *Vysokomol. Soed.* **35A**: 163–171.

360. Kwak, S-Y. and Kim, J-J. (1996) *Proceedings of the 1996 International Congress of Membranes and Membrane Processes* (ICOM'96), Yokokama, Japan, p. 162.

Index